Bayesian Evolutionary Analysis with BEAST

What are the models used in phylogenetic analysis and what exactly is involved in Bayesian evolutionary analysis using Markov chain Monte Carlo (MCMC) methods? How can you choose and apply these models, which parameterisations and priors make sense, and how can you diagnose Bayesian MCMC when things go wrong?

These are just a few of the questions answered in this comprehensive overview of Bayesian approaches to phylogenetics. From addressing theoretical aspects of the field to providing pragmatic advice on how to prepare and perform phylogenetic analysis, this practical guide also includes coverage of the interpretation of analyses and visualisation of phylogenies. The software architecture is described and a guide to developing BEAST 2.2 extensions is provided to allow these models to be extended further.

With an accompanying website (http://beast2.org/) providing example files and tutorials, this one-stop reference to applying the latest phylogenetic models in BEAST 2 will provide essential guidance for all users – from those using phylogenetic tools, to computational biologists and Bayesian statisticians.

Alexei J. Drummond is Professor of Computational Biology and Principal Investigator at the Allan Wilson Centre of Molecular Ecology and Evolution. He is the lead author of the BEAST software package and has gained a reputation in the field as one of the most knowledgeable experts for Bayesian evolutionary analyses.

Remco R. Bouckaert is a computer scientist with a strong background in Bayesian methods. He is the main architect of version 2 of BEAST and has been working on extensions to the BEAST software and other phylogenetics projects in Alexei Drummond's group at the University of Auckland.

Bayesian Evolutionary Analysis with BEAST

ALEXEI J. DRUMMOND

University of Auckland, New Zealand

REMCO R. BOUCKAERT

University of Auckland, New Zealand

CAMBRIDGE
UNIVERSITY PRESS

University Printing House, Cambridge CB2 8BS, United Kingdom

Cambridge University Press is part of the University of Cambridge.

It furthers the University's mission by disseminating knowledge in the pursuit of education, learning and research at the highest international levels of excellence.

www.cambridge.org
Information on this title: www.cambridge.org/9781107019652

First published 2015
Reprinted 2017

Printed in the United Kingdom by Clays, St Ives plc.

A catalogue record for this publication is available from the British Library

Library of Congress Cataloguing in Publication data
Drummond, Alexei J.
Bayesian evolutionary analysis ; with BEAST 2 / Alexei J. Drummond, University of Auckland, New Zealand, Remco R. Bouckaert, University of Auckland, New Zealand.
    pages   cm
Includes bibliographical references and index.
ISBN 978-1-107-01965-2 (Hardback)
1. Cladistic analysis–Data processing.   2. Bayesian statistical decision theory.   I. Bouckaert, Remco R.
II. Title.
QH83.D78 2015
578.01'2–dc23   2014044867

ISBN 978-1-107-01965-2 Hardback

Additional resources for this publication at http://beast2.org/book.html

Contents

Preface

This book consists of three parts: theory, practice and programming. The *theory part* covers theoretical background, which you need to get some insight in the various components of a phylogenetic analysis. This includes trees, substitution models, clock models and, of course, the machinery used in Bayesian analysis such as Markov chain Monte Carlo (MCMC) and Bayes factors.

In the *practice part* we start with a hands-on phylogenetic analysis and explain how to set up, run and interpret such an analysis. We examine various choices of prior, where each is appropriate, and how to use software such as BEAUti, FigTree and DensiTree to assist in a BEAST analysis. Special attention is paid to advanced analysis such as sampling from the prior, demographic reconstruction, phylogeography and inferring species trees from multilocus data. Interpreting the results of an analysis requires some care, as explained in the post-processing chapter, which has a section on troubleshooting with tips on detecting and preventing failures in MCMC analysis. A separate chapter is dedicated to visualising phylogenies.

BEAST 2.2 uses XML as a file format to specify various kinds of analysis. In the third part, the XML format and its design philosophy are described. BEAST 2.2 was developed as a platform for creating new Bayesian phylogenetic analysis methods, by a modular mechanism for extending the software. In the *programming part* we describe the software architecture and guide you through developing BEAST 2.2 extensions.

We recommend that everyone reads Part I for background information, especially introductory Chapter 1. Part II and Part III can be read independently. Users of BEAST should find much practical information in Part II, and may want to read about the XML format in Part III. Developers of new methods should read Part III, but will also find useful information about various methods in Part II.

The BEAST software can be downloaded from http://beast2.org and for developers, source code is available from https://github.com/CompEvol/beast2/. There is a lot of practical information available at the BEAST 2 wiki (http://beast2.org), including links to useful software such as Tracer and FigTree, a list of the latest packages and links to tutorials. The wiki is updated frequently. A BEAST users' group is accessible at http://groups.google.com/group/beast-users.

Acknowledgements

Many people made BEAST what it is today. Andrew Rambaut brought the first version of BEAST to fruition with AJD in the 'Oxford years' and has been one of the leaders of development ever since. Marc Suchard arrived on the scene a few years later, precipitating great advances in the software and methods, and continues to have a tremendous impact. All of the members of the core BEAST development team have been critical to the software's success.

Draft chapters of this book were greatly improved already by feedback from a large number of colleagues, including in alphabetical order, Richard Brown, David Bryant, Rampal Etienne, Alex Gavryushkin, Sasha Gavryushkina, Russell Gray, Simon Greenhill, Denise Kühnert, Tim Vaughan, David Welch, Walter Xie.

Paul O. Lewis created the idea for Figure 1.6. Section 2.3 is derived from work by Joseph Heled. Tanja Stadler co-wrote Chapter 2. Some material for Chapters 7 and 10 is derived from messages on the BEAST mailing list and the FAQ of the BEAST wiki. Section 8.4 is partly derived from 'A rough guide to SNAPP' (Bouckaert and Bryant 2012). Walter Xie was helpful in quality assurance of the software, in particular regression testing of BEAST ensuring that the analyses are valid. Parts of Chapter 4 derive from previous published work by AJD and co-authors.

The development of BEAST 2 was supported by four meetings funded by the National Evolutionary Synthesis Center (www.nescent.org). AJD was funded to write this book by a Rutherford Discovery Fellowship from the Royal Society of New Zealand.

Summary of most significant capabilities of BEAST 2

Analysis	Estimate phylogenies from alignments	
	Estimate dates of most recent common ancestors	
	Estimate gene and species trees	
	Infer population histories	
	Epidemic reconstruction	
	Estimate substitution rates	
	Phylogeography	
	Path sampling	
	Simulation studies	
Models	Trees	Gene trees, species trees, structured coalescent, serially sampled trees
	Tree-likelihood	Felsenstein, Threaded, BEAGLE
		Continuous, ancestral reconstruction
		SNAPP
		Auto partition
	Substitution models	JC96, HKY, TN93, GTR
		Covarion, stochastic Dollo
		RB, subst-BMA
		BLOSUM62, CPREV, Dayhoff, JTT, MTREV, WAG
	Frequency models	Fixed, estimated, empirical
	Site models	Gamma site model, mixture site model
	Tree priors	Coalescent constant, exponential, skyline
		Birth–death Yule, birth–death sampling skyline
		Yule with calibration correction
		Multispecies coalescent
	Clock models	Strict, relaxed, random local clock
	Prior distributions	Uniform, 1/X, normal, gamma, beta, etc.
Tools	BEAUti	GUI for specifying models
		Support for hierarchical models
		Flexible partition and parameter linking
		Read and write models
		Extensible through templates
		Manage BEAST packages
	BEAST	Run analysis specified by BEAUti
	ModelBuilder	GUI for visualising models
	LogCombiner	Tool for manipulating log files

	EBSPAnalyser	Reconstruct population history from EBSP analysis
	DensiTree	Tool for analysing tree distributions
	TreeAnnotator	Tool for creating summary trees from tree sets
	TreeSetAnalyser	Tool for calculating statistics on tree sets
	SequenceSimulator	Generate alignments for simulation studies
Check pointing	Resuming runs when ESS is not satisfactory Exchange partial states to reduce burn-in	
Documen-tation	Tutorials, Wiki, User forum This book	
Package support	Facilitate fast bug fixes and release cycles independent of core release cycle Package development independent of core releases	

Part I

Theory

1 Introduction

This book is part science, part technical, and all about the computational analysis of heritable traits: things like genes, languages, behaviours and morphology. This book is centred around the description of the theory and practice of a particular open source software package called BEAST (Bayesian evolutionary analysis by sampling trees). The BEAST software package started life as a small science project in New Zealand but it has since grown tremendously through the contributions of many scientists from around the world, chief among them the research groups of Alexei Drummond, Andrew Rambaut and Marc Suchard. A full list of contributors to the BEAST software package can be found on the BEAST GitHub page or printed to the screen when running the software.

Very few things challenge the imagination as much as does evolution. Every living thing is the result of the unfolding of this patient process. While the basic concepts of Darwinian evolution and natural selection are second nature to many of us, it is the detail of life's tapestry which still inspires an awe of the natural world. The scientific community has spent a couple of centuries trying to understand the intricacies of the evolutionary process, producing thousands of scientific articles on the subject. Despite this Herculean effort, it is tempting to say that we have only just scratched the surface.

As with many other fields of science, the study of biology has rapidly become dominated by the use of computers in recent years. Computers are the only way that biologists can effectively organise and analyse the vast amounts of genomic data that are now being collected by modern sequencing technologies. Although this revolution of data has really only just begun, it has already resulted in a flourishing industry of computer modelling of molecular evolution.

This book has the modest aim of describing this still new computational science of evolution, at least from the corner we are sitting in. In writing this book we have not aimed for it to be comprehensive and gladly admit that we mostly focus on the models that the BEAST software currently supports. Dealing, as we do, with computer models of evolution, there is a healthy dose of mathematics and statistics. However, we have made a great effort to describe in plain language, as clearly as we can, the essential concepts behind each of the models described in this book. We have also endeavoured to provide interesting examples to illustrate and introduce each of the models. We hope you enjoy it.

1.1   Molecular phylogenetics

The informational molecules central to all biology are deoxyribonucleic acid (DNA), ribonucleic acid (RNA) and protein sequences. These three classes of molecules are commonly referred to in the molecular evolutionary field as *molecular sequences*, and from a mathematical and computational point of view an informational molecule can often be treated simply as a linear sequence of symbols on a defined alphabet (see Figure 1.1). The individual building blocks of DNA and RNA are known as *nucleotides*, while proteins are composed of 20 different *amino acids*. For most life forms it is the DNA double-helix that stores the essential information underpinning the biological function of the organism and it is the (error-prone) replication of this DNA that transmits this information from one generation to the next. Given that replication is a binary reaction that starts with one genome and ends with two similar if not identical genomes, it is unsurprising that the natural and appropriate structure for visualising the replication process over multiple generations is a bifurcating tree. At the broadest scale of consideration the structure of this tree represents the relationships between species and higher-order taxonomic groups. But even when considering a single gene within a single species, the ancestral relationships among genes sampled from that species will be represented by a tree. Such trees have come to be referred to as *phylogenies* and it is becoming clear that the field of *molecular phylogenetics* is relevant to almost every scientific question that deals with the informational molecules of biology. Furthermore, many of the concepts developed to understand molecular evolution have turned out to transfer with little modification to the analysis of other types of heritable information in natural systems, including language and culture. It is unsurprising then that a book on computational evolutionary analysis would start with phylogenetics.

The study of phylogenetics is principally concerned with reconstructing the evolutionary history (phylogenetic tree) of related species, individuals or genes. Although algorithmic approaches to phylogenetics pre-date genetic data, it was the availability of genetic data, first allozymes and protein sequences, and then later DNA sequences, that provided the impetus for development in the area.

A phylogenetic tree is estimated from some data, typically a multiple sequence alignment (see Figure 1.2), representing a set of homologous (derived from a common ancestor) genes or genomic sequences that have been aligned, so that their comparable regions are matched up. The process of aligning a set of homologous sequences is itself a hard computational problem, and is in fact entangled with that of estimating a phylogenetic tree (Lunter et al. 2005; Redelings and Suchard 2005). Nevertheless, following convention we will – for the most part – assume that a multiple sequence alignment is known and predicate phylogenetic reconstruction on it.

DNA   {A,C,G,T}
RNA   {A,C,G,U}
Proteins {A,C,D,E,F,G,H,I,K,L,M,N,P,Q,R,S,T,V,W,Y}

Figure 1.1 The alphabets of the three informational molecular classes.

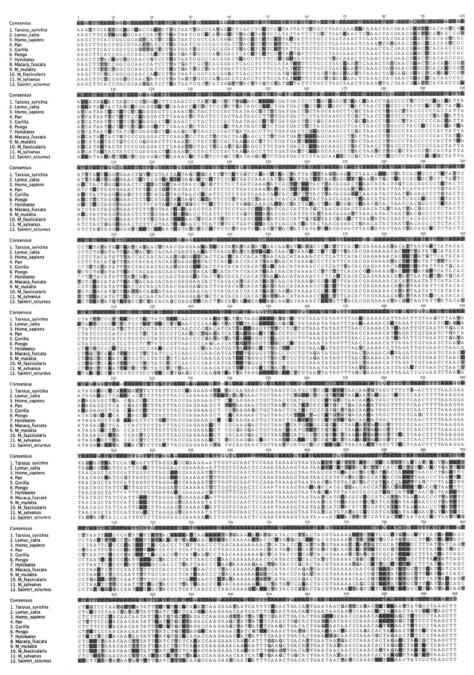

Figure 1.2 A small multiple sequence alignment of mitochondrial sequence fragments isolated from 12 species of primate. The alignment has 898 alignment columns and the individual sequences vary in length from 893 to 896 nucleotides long. Individual differences from the consensus sequence are highlighted. 373/898 (41.5%) sites are identical across all 12 species and the average pairwise identity is 75.7%. The data matrix size is 10 776 (898 × 12) with only 30 gap states. This represents a case in which obtaining an accurate multiple sequence alignment from unaligned sequences is quite easy and taking account of alignment uncertainty is probably unnecessary for most questions.

The statistical treatment of phylogenetics was made feasible by Felsenstein (1981), who described a computationally tractable approach to computing the probability of the sequence alignment given a phylogenetic tree and a model of molecular evolution, $\Pr\{D|T, \Omega\}$. This quantity is known as the *phylogenetic likelihood* of the tree and can be efficiently computed by the peeling algorithm (see Chapter 3). The statistical model of evolution that Felsenstein chose was a continuous-time Markov process (CTMP; see Section 3.1). A CTMP can be used to describe the evolution of a single nucleotide position in the genome, or for protein-coding sequences, either a codon position or the induced substitution process on the translated amino acids. By assuming that sites in a sequence alignment are evolving independently and identically a CTMP can be used to model the evolution of an entire multiple sequence alignment.

Although probabilistic modelling approaches to phylogenetics actually pre-date Sanger sequencing (Edwards and Cavalli-Sforza 1965), it was not until the last decade that probabilistic modelling became the dominant approach to phylogeny reconstruction (Felsenstein 2001). Part of that dominance has been due to the rise of Bayesian inference (Huelsenbeck et al. 2001), with its great flexibility in describing prior knowledge, its ability to be applied via the Metropolis–Hastings algorithm to complex highly parametric models, and the ease with which multiple sources of data can be integrated into a single analysis. The history of probabilistic models of molecular evolution and phylogenetics is a history of gradual refinement; a process of selection of those modelling variations that have the greatest utility in characterising the ever-growing empirical data. The utility of a new model has been evaluated either by how well it fits the data (formal model comparison or goodness-of-fit tests) or by the new questions that it allows a researcher to ask of the data.

1.2   Coalescent theory

When a gene tree has been estimated from individuals sampled from the same population, statistical properties of the tree can be used to learn about the population from which the sample was drawn. In particular the size of the population can be estimated using Kingman's *n-coalescent*, a stochastic model of gene genealogies described by Kingman (1982). *Coalescent theory* has developed greatly in the intervening decades and the resulting *genealogy-based population genetics* methods are routinely used to infer many fundamental parameters governing molecular evolution and population dynamics, including *effective population size* (Kuhner et al. 1995), rate of population growth or decline (Drummond et al. 2002; Kuhner et al. 1998), migration rates and population structure (Beerli and Felsenstein 1999, 2001; Ewing and Rodrigo 2006a; Ewing et al. 2004), recombination rates and reticulate ancestry (Bloomquist and Suchard 2010; Kuhner et al. 2000).

When the characteristic time scale of demographic fluctuations are comparable to the rate of accumulations of substitutions then past population dynamics are 'recorded' in the substitution patterns of molecular sequences.

The coalescent process is highly variable, so sampling multiple unlinked loci (Felsenstein 2006; Heled and Drummond 2008) or increasing the temporal spread of sampling times (Seo et al. 2002) can both be used to increase the statistical power of coalescent-based methods and improve the precision of estimates of both population size and substitution rate (Seo et al. 2002).

In many situations the precise functional form of the population size history is unknown, and simple population growth functions may not adequately describe the population history of interest. Non-parametric coalescent methods provide greater flexibility by estimating the population size as a function of time directly from the sequence data and can be used for data exploration to guide the choice of parametric population models for further analysis. These methods first cut the time-tree into segments, then estimate the population size of each segment separately according to the coalescent intervals within it.

Recently there has been renewed interest in developing mathematical modelling approaches and computational tools for investigating the interface between population processes and species phylogenies. The *multispecies coalescent* is a model of gene coalescence within a species tree (Figure 1.3; see Section 2.5.1 and Chapter 8 for further details). There is currently a large amount of development of phylogenetic inference techniques based on the multispecies coalescent model (Bryant et al. 2012; Heled and Drummond 2010; Liu et al. 2008, 2009a,b). Its many advantages over standard phylogenetic approaches centre around the ability to take into account real differences in the underlying gene tree among genes sampled from the same set of

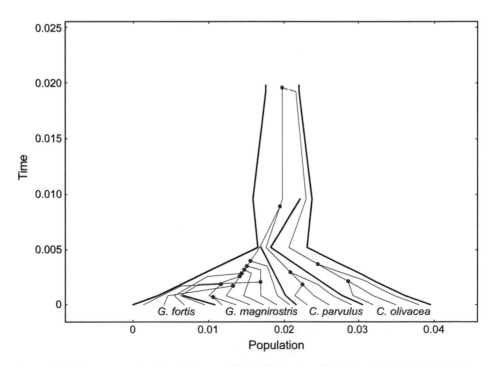

Figure 1.3 A four-taxa *species tree* with an embedded gene tree that relates multiple individuals from each of the sampled species. The species tree has (linear) population size functions associated with each branch, visually represented by the width of each species branch on the *x*-axis. The *y*-axis is a measure of time.

individuals from closely related species. Due to *incomplete lineage sorting* it is possible for unlinked genes from the same set of multispecies individuals to have different gene topologies, and for a particular gene to exhibit a *gene tree* that has different relationships among species than the true species tree. The multispecies coalescent can be employed to estimate the common species tree that best reconciles the coalescent-induced differences among genes, and provides more accurate estimates of divergence time and measures of topological uncertainty in the species tree. This exciting new field of coalescent-based species tree estimation is still in its infancy and there are many promising directions for development, including incorporation of population size changes (Heled and Drummond 2010), isolation with migration (Hey 2010), recombination and lateral gene transfer, among others.

1.3   Virus evolution and phylodynamics

A number of good recent reviews have been written about the impact of statistical phylogenetics and evolutionary analysis on the study of viral epidemiology (Kühnert et al. 2011; Pybus and Rambaut 2009; Volz et al. 2013). Although epidemic modelling of infectious diseases has a long history in both theoretical and empirical research, the term *phylodynamics* has a recent origin reflecting a move to integrate theory from mathematical epidemiology and statistical phylogenetics into a single predictive framework for studying viral evolutionary and epidemic dynamics. Many RNA viruses evolve so quickly that their evolution can be directly observed over months, years and decades (Drummond et al. 2003a). Figure 1.4 illustrates the effect that this has on the treatment of phylogenetic analysis.

Molecular phylogenetics has had a profound impact on the study of infectious diseases, particularly rapidly evolving infectious agents such as RNA viruses. It has given insight into the origins, evolutionary history, transmission routes and source populations of epidemic outbreaks and seasonal diseases. One of the key observations about rapidly evolving viruses is that the evolutionary and ecological processes occur on the same time scale (Pybus and Rambaut 2009). This is important for two reasons. First, it means that neutral genetic variation can track ecological processes and population dynamics, providing a record of past evolutionary events (e.g. genealogical relationships) and past ecological/population events (geographical spread and changes in population size and structure) that were not directly observed. Second, the concomitance of evolutionary and ecological processes leads to their interaction that, when non-trivial, necessitates joint analysis.

1.4   Before and beyond trees

Sequence alignment: After obtaining molecular sequences, a multiple sequence alignment must be performed before they can be analysed. Sequence alignment is a huge topic in itself (Durbin et al. 1998; Rosenberg 2009), and many techniques, including

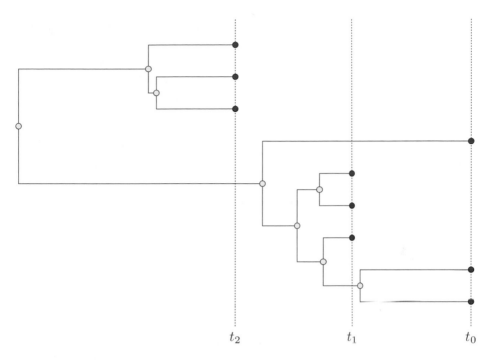

Figure 1.4 A hypothetical serially sampled gene tree of a rapidly evolving virus, showing that the sampling time interval ($\Delta t = t_2 - t_0$) represents a substantial fraction of the time back to the common ancestor. Red circles represent sampled viruses (three viruses sampled at each of three times) and yellow circles represent unsampled common ancestors.

dynamic programming, hidden Markov models and optimisation algorithms have been applied to this task. ClustalW and ClustalX (Larkin et al. 2007) are limited but widely used programs for this task. The Clustal algorithm uses a guide tree constructed by a distance-based method to progressively construct a multiple sequence alignment via pairwise alignments. Since most Bayesian phylogenetic analyses aim to reconstruct a tree, and the Clustal algorithm already assumes a tree to guide the alignment, it may be that a bias is introduced towards recovering the guide tree. This guide tree is based on a relatively simple model and it may contain errors resulting in sub-optimal alignments. T-Coffee (Notredame et al. 2000) is another popular program that builds a library of pairwise alignments to guide the construction of the complete alignment.

With larger data sets comes a need for high-throughput multiple sequence alignment algorithms; two popular and fast multiple sequence alignment algorithms are MUSCLE (Edgar 2004a,b) and MAFFT (Katoh and Standley 2013, 2014; Katoh et al. 2002)

However, a more principled approach in line with the philosophy of this book is to perform *statistical alignment* in which phylogenetic reconstructions and sequence alignments are simultaneously evaluated in a joint Bayesian analysis (Arunapuram et al. 2013; Bradley et al. 2009; Lunter et al. 2005; Novák et al. 2008; Redelings and Suchard 2005; Suchard and Redelings 2006). It has been shown that uncertainty in the alignment can lead to different conclusions (Wong et al. 2008), but in most cases it is hard to justify the extra computational effort required, and statistical alignment is not yet available in BEAST 2.2.

Ancestral recombination graphs: A phylogenetic tree is not always sufficient to reflect the sometimes complex evolutionary origin of a set of *homologous* gene sequences when processes such as *recombination*, *reassortment*, *gene duplication* or *lateral gene transfer* are involved in the evolutionary history.

Coalescent theory has been extended to account for recombination due to homologous crossover (Hudson 1990) and the *ancestral recombination graph* (ARG) (Bloomquist and Suchard 2010; Griffiths and Marjoram 1996; Kuhner 2006; Kuhner et al. 2000) is the combinatorial object that replaces a phylogenetic tree as the description of the ancestral (evolutionary) history. However, in this book we will limit ourselves to trees.

1.5　Probability and Bayesian inference

At the heart of this book is the idea that much of our understanding about molecular evolution and phylogeny will come from a characterisation of the results of random or stochastic processes. The sources of this randomness are varied, including the vagaries of chance that drive processes like mutation, birth, death and migration. An appropriate approach to modelling data that are generated by random processes is to consider the probabilities of various hypotheses given the observed data. In fact the concept of probability and the use of probability calculus within statistical inference procedures is pervasive in this book. We will not, however, attempt to introduce the concept of probability or inference in any formal way. We suggest (Bolstad 2011; Brooks et al. 2010; Gelman et al. 2004; Jaynes 2003; MacKay 2003) for a more thorough introduction or reminder about this fundamental material. In this section we will just lay out some of the terms, concepts and standard relationships and give a brief introduction to Bayesian inference.

1.5.1　A little probability theory

A random variable X represents a quantity whose value is uncertain and described by a probability distribution over all possible values. The set of possible values is called the sample space, denoted $\mathcal{S}_X$.

A probability distribution $\Pr(\cdot)$ on discrete mutually exclusive values, x in sample space $\mathcal{S}_X$ (i.e. $x \in \mathcal{S}_X$), sums to 1 over all values so that:

$$\sum_{x \in \mathcal{S}_X} \Pr(x) = 1,$$

and $0 \le \Pr(x) \le 1$ for all x. In this case we say X is discrete.

A classic example of $\mathcal{S}_X$ is the set of faces of a dice, $\mathcal{S}_X = \{\boxdot, \boxdot, \boxdot, \boxdot, \boxdot, \boxdot\}$, and for a random variable representing the outcome of rolling a fair dice, $\Pr(X = x) = 1/6$ for all $x \in \mathcal{S}_X$.

A conditional probability $\Pr(X = x | Y = y)$ gives the probability that random variable X takes on value x, given the condition that the random variable Y takes on value $y \in \mathcal{S}_Y$. This can be shortened to $\Pr(x | y)$ and the following relation exists:

$$\Pr(x) = \sum_{y \in \mathcal{S}_Y} \Pr(x | y) \Pr(y). \tag{1.1}$$

The joint probability of both $X = x$ *and* $Y = y$ occurring is

$$\Pr(x, y) = \Pr(x | y) \Pr(y).$$

This leads to another way to write Equation (1.1):

$$\Pr(x) = \sum_{y \in \mathcal{S}_Y} \Pr(x, y),$$

and $\Pr(x)$ is known as the marginal probability of $X = x$ in the context of the joint probability distribution on (X, Y).

A *probability density function* $f(x)$ defines a probability distribution over a continuous parameter $x \in \mathcal{S}_X$ (where $\mathcal{S}_X$ is now a continuous space) so that the integral of the density over the sample space sums to 1:

$$\int_{x \in \mathcal{S}_X} f(x) dx = 1,$$

and $f(x)$ is everywhere non-negative, but may take on values greater than 1 (see Figure 1.5 for an example). In this case the random variable X is continuous, and takes on a value in set $\mathcal{E} \subseteq \mathcal{S}_X$ with probability

$$\Pr(X \in \mathcal{E}) = \int_{x \in \mathcal{E}} f(x) dx.$$

This expresses the relationship between probabilities and probability densities on continuous random variables, where the left-hand side of the equation represents the probability of the continuous random variable X taking on a value in the set $\mathcal{E}$, and the right-hand side is the area under the density function in the set $\mathcal{E}$ (see Figure 1.5). If X is univariate real and $\mathcal{E}$ is interval $[a, b]$ then we have:

$$\Pr(a \leq X \leq b) = \int_a^b f(x) dx.$$

When x has probability density $f(\cdot)$ we write $X \sim f(\cdot)$. Finally, the expectation $E(\cdot)$ of a random variable $X \sim f(\cdot)$ can also be computed by an integral:

$$E(X) = \int_{x \in \mathcal{S}_X} x f(x) dx.$$

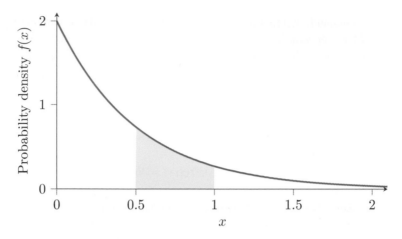

Figure 1.5 A probability density function $f(x) = 2e^{-2x}$ (i.e. an exponential distribution with a rate of 2). The area under the curve for $\Pr(0.5 \le X \le 1.0) = \int_{0.5}^{1.0} f(x)dx \approx 0.2325442$ is filled in.

1.5.2   Stochastic processes

A stochastic process is defined as a set of random variables $\{X_t\}$ indexed by a totally ordered set $t \in T$, which is a bit too general to be useful. In this book, we will mostly consider stochastic processes where the variables $\{X_t\}$ are discrete and t is continuous and represents time to describe models of evolution (Chapter 3) and birth–death processes (Chapter 2). Continuous state space and continuous time processes can be used in phylogeographic models (Chapter 5) and branch rate models (Chapter 4), but we will assume X_t to be discrete in the following. The relation between the stochastic variables is often described by a distribution $\Pr(X_{t=u}|X_{t\le s})$ with $u < s$. The distribution is usually associated with the name of the process, for example, when Pr is a Poisson distribution, we speak of a Poisson process. When $\Pr(X_{t=u}|X_{t\le s}) = \Pr(X_{t=u}|X_{t=s})$, we call the process Markovian. Most processes we deal with are CTMPs.

Stochastic processes cover a wide area requiring advanced mathematics. For further information on stochastic processes, Allen (2003) offers a good introduction. Stewart (1994) is useful for solving CTMPs numerically.

1.5.3   Bayesian inference

Bayesian inference is based on Bayes' formula:

$$p(\theta|D) = \frac{p(D|\theta)p(\theta)}{p(D)},$$

where $p(\cdot)$ could denote either a probability or a probability density, depending on whether θ and D are discrete or continuous. In a phylogenetic context, the data D are generally a discrete multiple sequence alignment and the parameters include continuous components (such as branch lengths and substitution rate parameters), so we have:

$$f(\theta|D) = \frac{\Pr(D|\theta)f(\theta)}{\Pr(D)}, \tag{1.2}$$

where D is the (discrete) sequence data, θ the set of (continuous) parameters in the model, including the tree, substitution model parameters, clock rates, *et cetera*.[1] $f(\theta|D)$ is called the *posterior distribution*, $\Pr(D|\theta)$ is the *likelihood*, $f(\theta)$ the *prior distribution* and $\Pr(D)$ the *marginal likelihood*. Informally, the reasoning starts with some prior belief about the model, encoded in the prior $f(\theta)$. After seeing some evidence in the form of data D, we update our belief about the model by multiplying the prior with the likelihood $\Pr(D|\theta)$, which is typically Felsenstein's phylogenetic likelihood. Our updated belief is the posterior $f(\theta|D)$. Specifying a prior distribution that represents prior knowledge about the model is not trivial and in fact a large part of this book deals with this topic.

As Equation (1.2) shows, we also need the marginal likelihood $\Pr(D)$ to complete the calculation. It is called the marginal likelihood since $\Pr(D)$ can be interpreted as the distribution $f(D|\theta)$ marginalised over θ, that is $\Pr(D) = \int_\theta f \Pr(D|\theta)f(\theta)\mathrm{d}\theta = \int_\theta f \Pr(D,\theta)\mathrm{d}\theta$. The marginal likelihood is almost never easy to compute, but fortunately, since it is constant during any analysis, the standard algorithm for sampling from the posterior does not require it (see Section 1.5.5 on Markov chain Monte Carlo below for details).

Bayesian inference fundamentally differs from many other methods of inference in that the outcome of an analysis is a probability distribution. For example, when inferring the age of the common ancestor of a set of taxa, instead of a point estimate, a Bayesian analysis results in a posterior distribution of ages. From this distribution, we can report, for example, the posterior mean and 95% highest posterior density (HPD) interval. The $x\%$ HPD interval is the smallest interval that includes $x\%$ of the posterior probability.

1.5.4 Non-informative and informative priors

An early tradition of Bayesian inference was to use *non-informative* priors. Such priors are designed to 'leave the data to tell the story' and are meant to represent the investigator's ignorance about the values of the parameters before the inference is performed. More recently this approach has been termed 'objective' Bayesian inference as opposed to the 'subjective' Bayesian inference performed when informative priors are used.

We will not attempt a careful appraisal of the differences between these two philosophies other than to make a few general remarks about when each of these approaches might be warranted.

It has been known since Laplace (1812) that if the constant prior $f(\theta) \propto 1$ is used, then – with small sample sizes – an inconsistency can occur where the result of a Bayesian analysis changes significantly based on the choice of the parameterisation

---

[1] We use $\Pr(\cdot)$ to denote probability distributions over discrete states, such as an alignment. Further, we use $f(\cdot)$ to denote densities over continuous spaces, such as rates or divergence times. Where the space is partly discrete and partly continuous, such as for (time-)trees, we will use the notation for densities.

(Berger and Bernardo 1992). A uniform prior in one parameterisation is typically not a uniform prior in another parameterisation. This problem was revisited in a phylogenetic context by Joseph Felsenstein in his book on phylogenetic inference (Felsenstein 2004), as an argument against using Bayesian inference for phylogenetics. Nevertheless, it has been suggested that this approach can sometimes be reasonable because the parameterisation is often chosen to reflect some notion of prior uniformity (Berger and Bernardo 1992). Furthermore, it was precisely the inconsistency under re-parameterisation that led Jeffreys (Jeffreys 1946, 1961) to develop his eponymous prior, one that is invariant to changes in parameterisation:

$$f(\theta) = \sqrt{\det(I(\theta))}, \tag{1.3}$$

where $I(\cdot)$ is the Fisher Information matrix of θ and $\det(\cdot)$ is the *determinant* of a matrix. This prior works well on one-dimensional problems, but its limitations for larger problems has led to the development of *reference priors* (Berger and Bernardo 1992).

Non-informative priors hold a natural appeal, especially for proponents of focusing exclusively on the evidence at hand. However, such priors can also often be *improper* in a formal sense if they don't have a finite integral. Such improper priors are extremely dangerous, and should be treated with great caution. For example, certain Bayesian methods (such as path sampling, see Section 1.5.7) require *proper priors* and will give undefined results if improper priors are used.

1.5.5   Markov chain Monte Carlo

In Bayesian inference we want to characterise the posterior distribution, which can be thought of as a multivariate function representing a convoluted landscape as shown in Figure 1.6. The figure shows just two dimensions, but since we are dealing with trees, the space has many more dimensions and is rather complex (see Chapter 11 for geometric descriptions of tree space). The Markov chain Monte Carlo (MCMC) algorithm (Hastings 1970; Metropolis et al. 1953) is an efficient way to explore the landscape. It provides a method of drawing samples from the posterior, that is, it samples values of the parameters $\theta_1, \theta_2, \theta_3, \ldots$ in proportion to their probability under the posterior distribution. These samples then form the basis of the study of the posterior such as calculating the mean. The application of MCMC to phylogenetic inference (Drummond and Rambaut 2007; Huelsenbeck and Ronquist 2001; Larget and Simon 1999; Lartillot et al. 2009; Lewis et al. 2005; Li et al. 2000; Mau and Newton 1997; Mau et al. 1999; Pagel and Meade 2004; Yang and Rannala 1997) and genealogical population genetics (Beerli 2006; Drummond et al. 2002; Ewing et al. 2004; Kuhner 2006; Vaughan et al. 2014; Wilson and Balding 1998) has received much attention.

The MCMC algorithm works as follows: as shown in Figure 1.6, the algorithm maintains a location in the landscape represented by an instantiation of the parameters θ. This is the current state. A new state θ' is proposed through a proposal distribution $q(\theta'|\theta)$, which typically favours a state that is close by, sometimes differing in the value of only

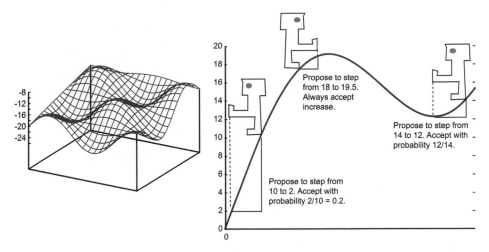

Figure 1.6 Left, posterior landscapes can contain many local optima. The MCMC sampler aims to return more samples from high posterior areas and fewer from low posterior regions. If run for sufficiently long, the sampler will visit all points. Right, the MCMC 'robot' evaluates a proposal location in the posterior landscape. If the proposed location is better, it accepts the proposal and moves to the new location. If the proposed location is worse, it will calculate the ratio of the posterior at the new location and that of the current location and accepts a step with probability equal to this ratio. So, if the proposed location is slightly worse, it will be accepted with high probability, but if the proposed location is much worse, it will almost never be accepted.

a single parameter. If the proposal distribution is symmetric, the new state is accepted with probability

$$\min\left(1, \frac{f(\theta'|D)}{f(\theta|D)}\right),\qquad(1.4)$$

hence if the posterior of the new state θ' is better it is always accepted, and if it is worse it is accepted by drawing a random number from the unit interval and if the number is less than the ratio $f(\theta'|D)/f(\theta|D)$ the new state is accepted. Either way, the step count of the chain is incremented.

A component of the proposal distribution may be termed an *operator*. Since a single operator typically only moves the state a small amount, the states θ and θ' will be highly dependent. However, after a sufficiently large number of steps, the states will become independent samples from the posterior. So, once every number of steps, the posterior and various attributes of the state are sampled and stored in a trace log and tree logs. Note that if the number of steps is too small, subsequent samples will still be autocorrelated, which means that the number of samples can be larger than the effective sample size (ESS).

It is a fine art to design and compose proposal distributions and the operators that implement them. The efficiency of MCMC algorithms crucially depends on a good mix of operators forming the proposal distribution. Note that a proposal distribution $q(\theta'|\theta)$ differs from the target distribution $f(\theta|D)$ in that changes to the former only affect the efficiency with which MCMC will produce an estimate of the latter. Some operators are not symmetric, so that $q(\theta'|\theta) \neq q(\theta|\theta')$. However, the MCMC algorithm described above assumes the probability of proposing θ' when in state θ is the same as the probability of proposing θ when in state θ'. The Metropolis–Hastings algorithm (Hastings 1970) is an MCMC algorithm that compensates to maintain reversibility by factoring in a Hastings ratio and accepts a proposed state with probability

$$\alpha = \min\left(1, \frac{f(\theta'|D)}{f(\theta|D)} \frac{q(\theta|\theta')}{q(\theta'|\theta)}\right). \tag{1.5}$$

The Hastings ratio corrects for any bias introduced by the proposal distribution. Correct calculation and implementation of non-symmetric operators in complex problems like phylogenetics is difficult (Holder et al. 2005).

Green (Green 1995; Green et al. 2003) describes a general method to calculate the Hastings ratio that also works when θ is not of fixed dimension. Green's recipe assumes that θ' can be reached from θ by selecting one or more random variables u, and likewise θ can be reached from θ' by selecting u' so that the vectors (θ', u') and (θ, u) have the same dimension. Let g and g' be the probability (densities) of selecting u and u' respectively. Green showed that the ratio $\frac{q(\theta|\theta')}{q(\theta'|\theta)}$ can be calculated as $\frac{g'(u')}{g(u)}|J|$ where $g(u)$ and $g'(u')$ is the density of u and u' respectively, and $|J|$ is the Jacobian of the tranformation $(\theta, u) \to (\theta', u')$.

For example, a scale operator with scale factor $\beta \in (0, 1)$ has a proposal distribution $q_\beta(\theta'|\theta)$ that transforms a parameter $\theta' = u\theta$ by selecting a random number u uniformly from the interval $(\beta, \frac{1}{\beta})$. Note that this scale operator has a probability of $(1-\beta)/(\frac{1}{\beta}-\beta)$ of decreasing and thus a higher probability of increasing θ, so the Hastings ratio cannot be 1. Moving between (θ, u) and (θ', u') such that $\theta' = u\theta$ requires that $u' = 1/u$. Consequently, the Hastings ratio when selecting u is

$$HR = \frac{g(u')}{g(u)} \left| \frac{\partial(\theta', u')}{\partial(\theta, u)} \right| = \frac{1}{\frac{1}{\beta} - \beta} \bigg/ \frac{1}{\frac{1}{\beta} - \beta} \left| \begin{array}{cc} \partial\theta'/\partial\theta & \partial\theta'/\partial u \\ \partial(1/u)/\partial\theta & \partial(1/u)/\partial u \end{array} \right|$$

$$= \left| \begin{array}{cc} u & \theta \\ 0 & \frac{1}{u^2} \end{array} \right| = \frac{1}{u}. \tag{1.6}$$

Operators typically only change a small part of the state, for instance only the clock rate while leaving the tree and substitution model parameters the same. Using operators that sample from the conditional distribution of a subset of parameters given the remaining parameters results in a Gibbs sampler (Geman and Geman 1984), and these operators can be very efficient in exploring the state space, but can be hard to implement since it requires that this conditional distribution is available.

Operators often have *tuning parameters* that influence how radical the proposals are. For the scale operator mentioned above, small values of β lead to bold moves, while values close to 1 lead to timid proposals. The value of β is set at the start of the run, but it can be tuned during the MCMC run so that it makes more bold moves if many proposals are accepted, or more timid if many are rejected. For example, in the BEAST inference engines (Bouckaert et al. 2014; Drummond et al. 2012), the way operator parameters are tuned is governed by an *operator schedule*, and there are a number of tuning functions.

$$\begin{bmatrix} - & r_1 & r_2 & r_3 \\ r_4 & - & r_5 & r_6 \\ r_7 & r_8 & - & r_9 \\ r_{10} & r_{11} & r_{12} & - \end{bmatrix} \begin{bmatrix} - & 1 & 0 & 0 \\ 1 & - & 1 & 0 \\ 0 & 1 & - & 1 \\ 0 & 0 & 1 & - \end{bmatrix} \begin{bmatrix} - & r_1 & 0 & 0 \\ r_4 & - & r_5 & 0 \\ 0 & r_8 & - & r_9 \\ 0 & 0 & r_{12} & - \end{bmatrix}$$

Figure 1.7 Left, rate matrix where all rates are continuously sampled. Middle, indicator matrix with binary values that are sampled. Right, rate matrix that is actually used, which combines rate from the rate matrix on the left with indicator variables in the middle.

Tuning typically only changes the parameters much during the start of the chain, and the tuning parameter will settle on a specific value as the chain progresses, guaranteeing the correctness of the resulting sample from the posterior distribution. In BEAST, not every operator has a tuning parameter, but if they do, its value will be reported at the end of a run, and suggest different values if the operator does not perform well.

It is quite common that the parameter space is not of a fixed dimension, for example when a nucleotide substitution model is chosen but the number of parameters is unknown. The MCMC algorithm can accommodate this using reversible jump (Green 1995) or Bayesian variable selection (BSVS, e.g. (Kuo and Mallick 1998; Wu and Drummond 2011; Wu et al. 2013)). With reversible jump, the state space actually changes dimension, which requires care in calculating the Hastings ratios of the operators that propose dimension changes, but is expected to be more computationally efficient than BSVS. BSVS involves a state space that contains the parameters of all the sub-models. A set of boolean indicator variables are also sampled that determine which model parameters are used and which are excluded from the likelihood calculation in each step of the Markov chain. Figure 1.7 shows an example for a rate matrix. The unused parameters are still part of the state space, and a prior is defined on them so proposals are still performed on these unused parameters, making BSVS less efficient than reversible jump. However, the benefit of BSVS is that it is easy to implement. EBSP (Heled and Drummond 2008) and discrete phylogeography (Lemey et al. 2009a) are examples that use BSVS, and the RB substitution model in the RBS plug-in is an example that uses reversible jump.

1.5.6 Bayesian model selection and model fit

The Bayesian approach to comparing theory M_1 with theory M_2 is to use a *Bayes factor* (BF), which is defined as the ratio of the marginal likelihoods. To compare the above models we would compute the marginal likelihoods $\Pr(D|M_1)$ and $\Pr(D|M_2)$, and calculate the BF as:

$$BF(M_1, M_2) = \frac{\Pr(D|M_1)}{\Pr(D|M_2)} \qquad (1.7)$$

for data D. Although it has been suppressed in previous sections, here we make the posterior dependence on the model explicit by including the model on the right of vertical bar. So $\Pr(D|M)$ is the marginal likelihood with respect to the prior on the

Table 1.1 Interpreting Bayes factors

BF range	ln(BF) range	$\log_{10}$(BF) range	Interpretation
1–3	0–1.1	0–0.5	Hardly worth mentioning
3–20	1.1–3	0.5–1.3	Positive support
20–150	3–5	1.3–2.2	Strong support
>150	>5	>2.2	Overwhelming support

parameters θ_M of model M:

$$\Pr(D|M) = \int_{\theta_M} \Pr(D|\theta_M, M) f(\theta_M|M) d\theta_M, \tag{1.8}$$

where $\Pr(D|\theta_M, M)$ is the likelihood (represented by $\Pr(D|\theta)$ in the previous section) with the dependence on model M made explicit. There are a number of ways of approximately calculating the marginal likelihood of each model, and therefore the BF between them, that can be achieved by processing the output of two Bayesian MCMC analyses. A simple method first described by Newton and Raftery (1994) computes the BF via *importance sampling* with the posterior as the importance distribution. With this importance distribution it turns out that the harmonic mean of the posterior sample of likelihoods is an estimator of the marginal likelihood. By calculating this harmonic mean estimator (HME) from the posterior output of each of the competing models and then taking the difference (of the estimated log marginal likelihoods) one can obtain an estimate of the log BF and decide whether the BF (log BF) is large enough to strongly favour one model over the other. Interpretation of the BF according to Kass and Raftery (1995) is shown in Table 1.1. Note in log space $\ln(20) \approx 3$, so a log BF over 3 is strong support for the favoured model. Confusingly, both BF and log BF are called Bayes factor so one has to be aware which version is used.

Although easy to calculate, the HME is known to be a very poor estimator of marginal likelihood (e.g. Baele et al. 2012, 2013a). An alternative estimator that is as convenient to calculate as HME, but slightly more reliable, is based on Akaike's information criterion (AIC; Akaike 1974) through MCMC (AICM; Raftery et al. 2007). AICM is based on the observation that the distribution of the log-likelihood approaches a gamma distribution for a sufficiently large amount of data. AICM directly estimates the parameters of a gamma distribution from the log-likelihoods in an MCMC sample. While AICM can be more reliable than HME, they both tend to be poor estimators of marginal likelihood compared to the more computationally intensive path sampling (see Section 1.5.7).

However calculated, it is important to realise that BFs offer only a relative comparison of explanatory power, dealing as they do with the relative probability of the data under competing models. It could be that although one model is much better than the other, both models are very poor explanations of the data. As a result, absolute goodness-of-fit tests, such as posterior predictive simulation tests, are gaining popularity for Bayesian model selection and model fit (Brown 2014; Drummond and Suchard 2008).

1.5.7 Path sampling

Path sampling and the stepping stone algorithm (Baele et al. 2012; Xie et al. 2011) are techniques to estimate the marginal likelihood by running an MCMC chain that samples from $f(\theta)\Pr(D|\theta)^\beta$ for a schedule of different values of β. The difference between the path sampling and stepping stone algorithms lies in the details of how the marginal likelihood estimator is constructed once the empirical likelihoods for each of the β values have been obtained.

Path sampling can also be used to calculate the BF between competing models directly using an MCMC chain that samples from:

$$\left[f(\theta_{M_1}|M_1)\Pr(D|\theta_{M_1},M_1)\right]^\beta \left[f(\theta_{M_2}|M_2)\Pr(D|\theta_{M_2},M_2)\right]^{1-\beta} \qquad (1.9)$$

where β runs from 0 to 1 following a sigmoid function (Baele et al. 2013). Such direct comparison of models instead of estimating marginal likelihoods in separate path sampling analyses has the advantage that fewer steps are required. Furthermore, computational gains can be made if large parts of the model are shared by models M_1 and M_2. For example, if M_1 and M_2 only differ in the tree prior, but the likelihoods are the same, $\Pr(D|\theta_{M_1},M_1) = \Pr(D|\theta_{M_2},M_2)$, Equation (1.9) reduces to

$$f(\theta_{M_1}|M_1)^\beta f(\theta_{M_2}|M_2)^{1-\beta} \Pr(D|\theta_{M_1},M_1),$$

and the likelihood needs to be calculated only once.

1.5.8 The appeal of Bayesian analysis

After this short introduction to Bayesian methods, one may wonder what the appeal is over maximum likelihood (ML) and parsimony-based methods in phylogenetic analysis. There are both theoretical and practical advantages. Arguably, the conflict between probabilistic-based methods and cladistic parsimony as chronicled by Felsenstein (2001) has resulted in a split into two different schools where the probabilists use model-based methods and the others do not. Without models it is hard to quantify the uncertainty in an analysis, which is a requisite for doing science.

Maximum likelihood methods find the model parameter values that maximise the probability of the data D, that is $\max_\theta \Pr(D|\theta)$. Bayesian inference finds the posterior distribution $f(\theta|D)$. Firstly, this implies that through the Bayesian method we can infer generalisations about the model represented by θ, while in ML this cannot be justified (Jaynes 2003). Secondly, since Bayesian methods return a distribution instead of a single set of parameter values it is easy to answer questions of interest, such as what is the probability that the root height of a tree lies in a given range.

The Bayesian method requires specifying priors, which is a mixed blessing since it requires extra effort and care, but also allows constraining the analysis when information of, say, the substitution rate is available from independent sources or from the literature. Furthermore, the priors have to be made explicit, whereas in ML analysis a bias is easily hidden. For example, when a substitution rate is allowed to uniformly range from 0 to 1, we implicitly express that we favour with 99.9% probability that the rate is higher than

0.001, even though we know that only a few fast-evolving viruses have such high rates (Duffy et al. 2008).

Priors can be set using the objective view, which is based on the principle that priors should be non-informative and only the model and data should influence the posterior (Berger 2006). Alternatively, subjective Bayesian practitioners set priors so as to contain prior information from previous experience, expert opinions and information from the literature (Goldstein 2006). This offers the benefit from a more pragmatic point of view that the Bayesian approach allows us to combine information from heterogeneous sources into a single analysis based on a formal foundation in probability theory. For example, integrating DNA sequence data, information about geographic distribution and data from the fossil record into a single analysis becomes quite straightforward, as we shall see later in this book.

Another practical consideration is that after setting up priors and data, an MCMC algorithm does not generally require as much special attention or tuning as many hill-climbing or simulated annealing algorithms used in ML. A well-designed MCMC algorithm just runs, automatically tunes itself and produces a posterior distribution. The MCMC algorithm is guaranteed to converge in the limit of the number of samples (Hastings 1970; Metropolis et al. 1953), but in practice it tends to converge much faster. The algorithm is particularly suited to navigating the multi-modal and highly peaked probability landscape typical of phylogenetic problems.

2 Evolutionary trees

Alexei Drummond and Tanja Stadler

2.1 Types of trees

This book is about evolution, and one of the fundamental features of evolutionary analysis is the tree. The terms tree and phylogeny are used quite loosely in the literature for the purposes of describing a number of quite distinct objects. *Evolutionary trees* are a subset of the group of objects that graph theorists know as trees, which are themselves connected graphs that do not contain cycles. An evolutionary tree typically has labelled leaf nodes (tips) and unlabelled internal nodes (an internal node may also be known as a divergence or coalescence). The leaf nodes are labelled with taxa, which might represent an individual organism, or a whole species, or more typically just a gene fragment, while the internal nodes represent unsampled (and thus inferred) common ancestors of the sampled taxa. For reasons mainly of history, the types of trees are many and varied; in the following we introduce the main types.

2.1.1 Rooted and unrooted trees

One of the more important distinctions is between rooted trees and unrooted trees (Figure 2.1). Both of the trees in Figure 2.1 describe the evolutionary relationships between four taxa labelled A to D.

A rooted tree has a notion of the direction in which evolution occurred. One internal node is identified as the root, and evolution proceeds from the root to the leaves. A tree is said to be binary if its internal nodes always have precisely two children. A rooted binary tree of n taxa can be described by $2n - 1$ nodes and $2n - 2$ branches, each with an associated branch length. A rooted binary tree is displayed in Figure 2.1a; note that the path length between two leaf nodes should be measured only by the sum of the lengths of the horizontal lines along the shortest path connecting the two leaves. The vertical lines exist purely for the purpose of visual layout. Taxon A is actually the shortest distance of all the taxa to taxon D, even though taxon A is the furthest from D vertically.

In contrast, an unrooted tree does not have a root and so does not admit any knowledge of which direction evolution 'flows'. The starting point is not known, so the tree is generally drawn with the leaf nodes spread around the perimeter of the diagram. We

Tanja Stadler, ETH Zürich, Department of Biosystems Science & Engineering (D-BSSE), Mattenstrasse 26, 4058 Basel, Switzerland

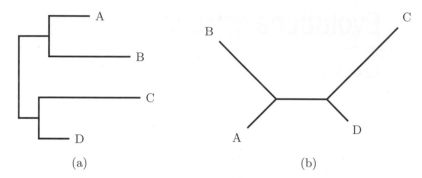

Figure 2.1 Two leaf-labelled binary trees; (a) is rooted and (b) is unrooted.

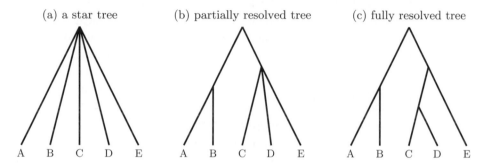

Figure 2.2 Multifurcating and bifurcating trees.

know that the evolutionary process finishes at the leaves, but we do not know from which point in the tree it starts. An unrooted binary tree is a tree in which each node has one branch (leaf) or three branches (internal node) attached, and can be obtained from a rooted binary tree through replacing the root node and its two attached branches by a single branch. An unrooted binary tree of n taxa can be described by $2n - 2$ nodes and $2n - 3$ branches with branch lengths. In the unrooted tree diagram in Figure 2.1b the path length between two leaf nodes is simply the sum of the branch lengths along the shortest path connecting the two leaves. This figure makes it more obvious that taxon A is the closest to taxon D.

There are many different units that the branch lengths of a tree could be expressed in, but a common unit that is used for trees estimated from molecular data is substitutions per site. We will expand more on this later in the book when we examine how one estimates a tree using real molecular sequence data.

2.1.2   Multifurcating trees and polytomies

A polytomy in a rooted tree is an internal node that has more than two children. Multifurcating trees (as opposed to bifurcating/binary trees) are those that have one or more polytomies. Polytomies are sometimes used to represent a lack of knowledge about the true relationships in some part of a tree. In this context the tree is sometimes called partially resolved. A fully resolved tree is a binary tree by another name. Figure 2.2

illustrates a completely unresolved (star) tree, along with a partially resolved (and thus multifurcating) tree and a fully resolved (and thus binary or bifurcating) tree.

2.1.3 Time-trees

A time-tree is a special form of rooted tree, namely a rooted tree where branch lengths correspond to calendar time, i.e. the duration between branching events. A time-tree of n taxa can be described by $2n - 2$ edges (branches) and $2n - 1$ nodes with associated node times (note that assigning $2n - 2$ branch lengths and the time of one node is equivalent to assigning $2n - 1$ node times). The times of the internal nodes are called divergence times, ages or coalescent times, while the times of the leaves are known as sampling times. Figure 2.5 is an example of a time-tree. Often we are interested in trees in which all taxa are represented by present-day samples, such that all the sampling times are the same. In this case it is common for the sampling times to be set to zero, and the divergence times to be specified as times before present (ages), so that time increases into the past.

2.1.4 BEAST infers time-trees

In unrooted trees, branch lengths typically represent the amount of evolutionary change, while in rooted trees, branch lengths represent either amount or duration of evolution (and we call the rooted tree which represents duration of evolution the time-tree). BEAST infers both types of rooted trees. Unrooted trees can be estimated in programs such as MrBayes (Ronquist and Huelsenbeck 2003) or PhyloBayes. When inferring rooted trees, we link the amount and duration of evolution. If amount and duration of evolution is the same, we assume a strict clock model; however, when there is

Table 2.1 Notation used for time-trees in all subsequent sections and chapters

Notation	Description
n	The number of taxa (and therefore leaf nodes) in the tree
I	The set of leaf nodes in a ranked tree. $|I| = n$
Y	The set of internal nodes in a ranked tree. There is a total order on internal nodes so that $x > y$ for $x, y \in Y$ implies that node x is closer to the root than node y. $|Y| = n - 1$, $I \cap Y = \emptyset$
V	The set of nodes in a ranked tree: $|V| = 2n - 1$, $V = I \cup Y$
$\langle i, j \rangle$	An ordered edge in a ranked tree such that i is the parent of j. $i > j$, $i \in Y$, $j \in V$
R	A set of ordered edges $\langle i, j \rangle$ representing a ranked tree. $|R| = 2n - 2$. $R \in \mathcal{R}_n$
t_i	The time of node i, $i \in V$. $i > j$ implies $t_i > t_j$.
τ_i	The length of the time interval that a time-tree of contemporaneous samples has i lineages
$\mathbf{t} = \{t_i : i \in V\}$	The set of all node times in a time-tree. $|\mathbf{t}| = |V|$
$g = \langle R, \mathbf{t} \rangle$	A time-tree consisting of a ranked tree R and associated node times $\mathbf{t}$.
$\mathcal{R}_n$	The set of all ranked trees of size n.
$\mathcal{G}_n$	The set of all time-trees of size n.
$\mathcal{G}_S$	The set of time-trees with at least one sample, that is, $\mathcal{G}_S = (\mathcal{G}_1 \cup \mathcal{G}_2 \cup \mathcal{G}_3 \ldots)$

sufficiently large variance in the relation between amount and duration of evolution, a relaxed clock model needs to be considered (see Chapter 4) since it attempts to model this variance. Since in a time-tree node heights correspond to the ages of the nodes (or at least relative ages if there is no calibration information), such rooted tree models have fewer parameters than unrooted tree models have, approaching roughly half for large number of taxa (for n taxa, there are $2n - 3$ branch lengths for unrooted trees, while for rooted time trees there are $n - 1$ node heights for internal nodes and a constant but low number of parameters for the clock model). As BEAST infers rooted, binary (time) trees we will focus on these objects. If not specified otherwise, a tree will refer to a rooted binary tree in the following.

2.2　Counting trees

Estimating a tree from molecular data turns out to be a difficult problem. The difficulty of the problem can be appreciated when one considers how many possible tree topologies, i.e. trees without branch lengths, there are. Consider $\mathcal{T}_n$, the set of all tip-labelled rooted binary trees of n taxa. The number of distinct tip-labelled rooted binary trees $|\mathcal{T}_n|$ for n taxa is (Cavalli-Sforza and Edwards 1967):

$$|\mathcal{T}_n| = \prod_{k=2}^{n}(2k - 3) = \frac{(2n - 3)!}{2^{n-2}(n - 2)!}. \tag{2.1}$$

Table 2.2 shows the number of tip-labelled rooted trees up to ten taxa, and other related quantities.

2.2.1　Tree shapes

An *unlabelled* tree is sometimes known as a *tree shape*. For three taxa there is only one rooted binary tree shape: $\bigwedge$, while for four taxa there are two shapes: the comb tree, $\bigwedge\!\!\bigwedge$, and the balanced tree, $\bigwedge\ \bigwedge$.

Table 2.2 The number of unlabelled rooted tree shapes, the number of labelled rooted trees, the number of labelled ranked trees (on contemporaneous tips) and the number of fully ranked trees (on distinctly timed tips) as a function of the number of taxa, n

n	#shapes	#trees, $|\mathcal{T}_n|$	#ranked trees, $|\mathcal{R}_n|$	#fully ranked trees, $|\mathcal{F}_n|$
2	1	1	1	1
3	1	3	3	4
4	2	15	18	34
5	3	105	180	496
6	6	945	2700	11 056
7	11	10 395	56 700	349 504
8	23	135 135	1 587 600	14 873 104
9	46	2 027 025	57 153 600	819 786 496
10	98	34 459 425	2 571 912 000	56 814 228 736

In general, the number of tree shapes (or unlabelled rooted tree topologies) of n taxa (a_n) is given by (Cavalli-Sforza and Edwards 1967):

$$a_n = \begin{cases} \sum_{i=1}^{(n-1)/2} a_i a_{n-i} & n \text{ is odd} \\ a_1 a_{n-1} + a_2 a_{n-2} + \cdots + \frac{1}{2} a_{n/2}(a_{n/2} + 1) & n \text{ is even}. \end{cases} \quad (2.2)$$

This result can easily be obtained by considering that for any tree shape of size n, it must be composed of two smaller tree shapes of size i and $n - i$ that are joined by the root to make the larger tree shape. This leads directly to the simple recursion above. So for five taxa, the two branches descending from the root can split the taxa into subtrees of size (1 and 4) or (2 and 3). There are $a_1 a_4 = 2$ tree shapes of the first kind and $a_2 a_3 = 1$ tree shape of the second kind, giving $a_5 = a_1 a_4 + a_2 a_3 = 3$. This result can be built upon to obtain a_6 and so on.

2.2.2 Ranked trees

A ranked tree is a rooted binary tree topology, where in addition the order of the internal nodes is given. There are more ranked trees than rooted tree topologies, and they are important because many natural tree priors are uniform on ranked trees rather than tree shapes. The number of ranked trees of n *contemporaneous* taxa, $F(n) = |\mathcal{R}_n|$, is:

$$F(n) = |\mathcal{R}_n| = \prod_{k=2}^{n} \binom{k}{2} = \frac{n!(n-1)!}{2^{n-1}}. \quad (2.3)$$

All ranked trees with four tips are shown in Figure 2.3. When a tree has non-contemporaneous times for the sampled taxa we term the tree *fully ranked* (Gavryushkina et al. 2013) and the number of fully ranked trees of n tips can be computed by recursion:

$$F(n_1, \ldots, n_m) = \sum_{i=1}^{n_m} \frac{|\mathcal{R}_{n_m}|}{|\mathcal{R}_i|} F(n_1, n_2, \ldots n_{m-2}, n_{m-1} + i), \quad (2.4)$$

where n_i is the number of tips in the ith set of tips, grouped by sample time (see Gavryushkina et al. 2013 for details).

Let $\mathcal{F}_n$ be the set of fully ranked trees on n tips, each with a distinct sampling time, then:

$$|\mathcal{F}_n| = F(\overbrace{1, 1, \cdots, 1}^{n \text{ times}}).$$

Table 2.2 shows how $|\mathcal{F}_n|$ grows with n.

2.2.3 Time-trees

Consider $\mathcal{G}_n$, the (infinite) set of time-trees of size n. $\mathcal{G}_n$ can be constructed by the Cartesian product of (1) the set of ranked trees $\mathcal{R}_n$ and (2) **D** the set of ordered divergence

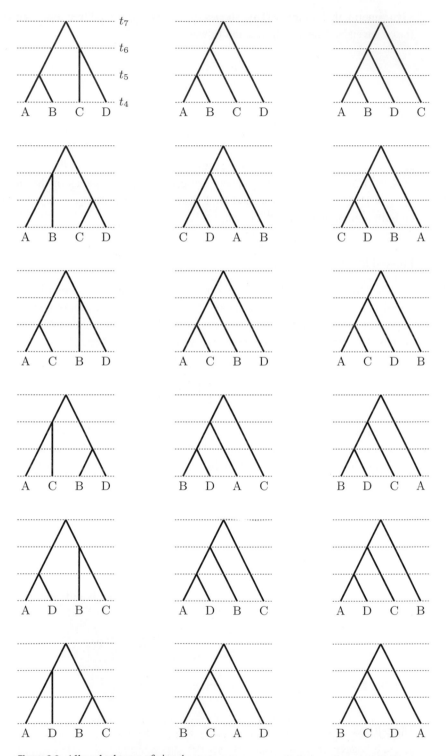

Figure 2.3 All ranked trees of size 4.

sampling times, $\mathbf{t} = \{t_1 = 0, t_2, \ldots, t_{2n-1}\}, t_k \geq t_{k-1}, \mathbf{t} \in \mathbf{D}$ (with time increasing into the past):

$$\mathcal{G}_n = \mathcal{R}_n \times \mathbf{D} = \{(R, \mathbf{t}) | R \in \mathcal{R}_n, \mathbf{t} \in \mathbf{D}\}.$$

In the remainder of this chapter, we discuss models giving rise to time-trees (and thus also to ranked trees and tree shapes).

2.3 The coalescent

Much of theoretical population genetics is based on the idealised Wright–Fisher model which gives rise, as we explain below, to a distribution of time-trees for large population sizes N, which is the coalescent tree distribution (see for example Hein et al. 2004 for a primer on coalescent theory).

The Wright–Fisher model in its simplest form assumes (1) constant population size N, (2) discrete generations, (3) complete mixing.

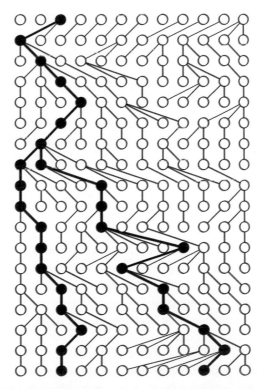

Figure 2.4 A haploid Wright–Fisher population of a dozen individuals with the ancestry of two individuals sampled from the current generation traced back in time. Going back in time, the traced lineages coalesce on a common ancestor 11 generations in the past, and from there onwards the ancestry is shared.

Now consider two random members of the current generation from a population of fixed size N (refer to Figure 2.4). By complete mixing, the probability they share a *concestor* (common ancestor) in the previous generation is $1/N$. It can easily be shown by a recursive argument that the probability the concestor is t generations back is

$$\Pr\{t\} = \frac{1}{N} \left(1 - \frac{1}{N}\right)^{t-1}.$$

It follows that $X = t - 1$ has a geometric distribution with a success rate of $\lambda = 1/N$, and so has mean N and variance of $N^2 - N \approx N^2$.

With k lineages the time to the first coalescence can be derived in the same way. The probability that none of the k lineages coalesces in the previous generation is

$$\left(\frac{N-1}{N}\right)\left(\frac{N-2}{N}\right)\cdots\left(\frac{N-k+1}{N}\right) = 1 - \frac{\binom{k}{2}}{N} + O(1/N^2).$$

Thus the probability of a coalescent event is $\binom{k}{2}/N + O(1/N^2)$. Now for large N we can drop the order $O(N^{-2})$ term, and this results in a success rate of $\lambda = \binom{k}{2}/N$ and the mean waiting time to the first coalescence among k lineages (τ_k) of

$$E(\tau_k) = \frac{N}{\binom{k}{2}}.$$

Dropping $O(N^{-2})$ implicitly assumes that N is much larger than k such that two coalescent events occurring in the same generation is negligible.

Kingman (1982) showed that as N grows the coalescent process converges to a continuous-time Markov chain. The rate of coalescence in the Markov chain is $\lambda = \binom{k}{2}/N$, i.e. going back in time, the probability of a pair coalescing from k lineages on a short time interval Δt is $O(\lambda \Delta t)$. Unsurprisingly the solution turns out to be the exponential distribution:

$$f(\tau_k) = \frac{\binom{k}{2}}{N} \exp\left(-\frac{\binom{k}{2}\tau_k}{N}\right).$$

Applied to a sample genealogy, Kingman's coalescent (Kingman 1982) describes the distribution of coalescent times in the genealogy as a function of the size of the population from which it was drawn, assuming an idealised Wright–Fisher population (Fisher 1930; Wright 1931). The coalescent can also be obtained by taking the limit in large N from the continuous-time finite population Moran model (Moran 1958, 1962).

In practice the idealised assumptions underlying the coalescent are often not met perfectly in real data. Common features of real populations that aren't explicitly taken into account in the idealised formulation of coalescent theory include variation in reproductive success, population age structure and unequal sex ratios (Wright 1931). These extra complexities mean that the estimated population size parameter is almost always smaller than the actual census size of a population (Wright 1931). It is therefore common to term the parameter estimated using coalescent theory the *effective population size* and denote it N_e. One way to interpret coalescent N_e is to say that the natural population exhibits sample genealogies that have the statistical properties of an idealised population of size N_e. Although this means that the absolute values of N_e are difficult to relate to the true census size, it allows different populations to be compared on a common scale (Sjödin et al. 2005). Theoretical extensions to the concept of *coalescent effective population size* have also received recent attention (Wakeley and Sargsyan 2009) and the complexities of interpreting effective population size in the context of HIV-1 evolution has received much consideration (Kouyos et al. 2006). However, see Gillespie (2001) for an argument that these neutral evolution models are irrelevant to much real data because neutral loci will frequently be sufficiently close to loci under selection, that *genetic draft* and *genetic hitchhiking* will destroy the relationship between population size and genetic diversity that coalescent theory relies on for its inferential power.

The original formulation was for a constant population (as outlined above), but the theory has been generalised in a number of ways (see Hudson (1990) for a classic review), including its application to recombination (Hudson 1987; Hudson and Kaplan 1985), island migration models (Hudson 1990; Slatkin 1991; Tajima 1989), population divergence (Tajima 1983; Takahata 1989) and deterministically varying functions of population size for which the integral $\int_{t_0}^{t_1} N(t)^{-1} dt$ can be computed (Griffiths and Tavaré 1994). Since all but the final extension requires more complex combinatorial objects than time-trees (either *ancestral recombination graphs* or *structured time-trees* in which tips and internal branch segments are discriminated by their subpopulation), we will largely restrict ourselves to single population non-recombining coalescent models in the following sections. For information about structured time-tree models, see Chapter 5.

2.3.1   Coalescent with changing population size in a well-mixed population

Parametric models with a pre-defined population function, such as exponential growth, expansion model and logistic growth models can easily be used in a coalescent framework. The logistic population model was one of the first non-trivial population models employed in a coalescent framework (Pybus et al. 2001) and analytic solutions for the coalescent likelihood for a number of parametric models were implemented in the package GENIE (Pybus and Rambaut 2002). Fully Bayesian inference under these models followed soon after. For example a 'piecewise-logistic' population model was employed in a Bayesian coalescent framework to estimate the population history of hepatitis C virus (HCV) genotype 4a infections in Egypt (Pybus et al. 2003). This analysis demonstrated a rapid expansion of HCV in Egypt between 1930 and 1955, consistent with the hypothesis that public health campaigns to administer anti-schistosomiasis injections had caused the expansion of an HCV epidemic in Egypt through usage of HCV-contaminated needles.

The integration of likelihood-based phylogenetic methods and population genetics through the coalescent has provided fertile ground for new developments. Many coalescent-based estimation methods focus on a single genealogy (Felsenstein 1992; Fu 1994; Nee et al. 1995; Pybus et al. 2000) that is typically obtained using standard phylogenetic reconstruction methods. For example, a maximum likelihood tree (under

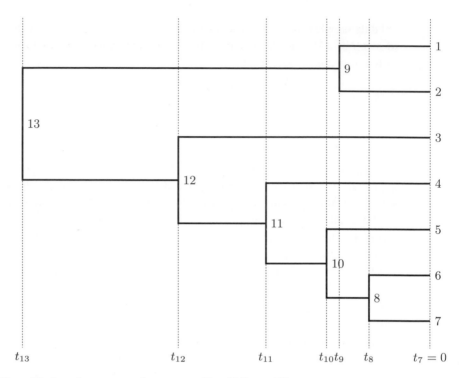

Figure 2.5 A coalescent tree of seven taxa. $Y = \{8, 9, \dots, 13\}$.

clock constraints) can be obtained and then used to obtain a maximum likelihood estimate of the mutation-scaled effective population size $N_e\mu$ (where μ is mutation rate per generation time) using coalescent theory. If the mutation rate per generation is known, then N_e can be estimated directly from a time-tree in which the time is expressed in generations; otherwise if the mutation rate is known in some calendar units, μ_c, then the estimated population size parameter will be $N_e g_c$ where $\mu = \mu_c g_c$ and thus g_c is the generation time in calendar units.

Consider a demographic function $N_e(x)$ and an ordered set of node times $\mathbf{t} = \{t_i : i \in V\}$. Let k_i denote the number of lineages co-existing in the time interval (t_{i-1}, t_i) between node $i - 1$ and node i. Note that for a contemporaneous tree, k_i decreases monotonically with increasing i.

The probability those times are the result of the coalescent process reducing n lineages into one is obtained by multiplying the (independent) probabilities for each coalescence event,

$$f(\mathbf{t}|N_e(x)) = \prod_{i \in Y} \frac{\binom{k_i}{2}}{N_e(t_i)} \prod_{i \in V} \exp\left(-\int_{t_{i-1}}^{t_i} \frac{\binom{k_i}{2}}{N_e(x)}\, dx\right), \tag{2.5}$$

where $t_0 = 0$ is defined to effect compact notation in the second product. Note that the second product is over all nodes (including leaf nodes) to provide for generality of the result when leaf nodes are non-contemporaneous (i.e. dated tips).

For a time-tree $g = \langle R, \mathbf{t} \rangle$ with *contemporaneous* tips, composed of a ranked tree topology $R \in \mathcal{R}_n$ and coalescent times $\mathbf{t}$, the probability density becomes:

$$f(g|N_e(x)) = \frac{1}{|\mathcal{R}_n|} f(\mathbf{t}|N_e(x)),$$

since there are $|\mathcal{R}_n|$ ranked trees of equal probability under the coalescent process (Aldous 2001). *Note that this second result only holds for contemporaneous tips, as not all ranked trees are equally probable under the coalescent when tips are non-contemporaneous* (see Section 2.3.2 for non-contemporaneous tips). The function $N_e(x)$ that maximises the likelihood of the time-tree g is the maximum likelihood estimate of the population size history. For the simplest case of constant population size $N_e(x) = N_e$ and contemporaneous tips, this density becomes:

$$f(g|N_e) = \prod_{i \in Y} \frac{1}{N_e} \exp\left(-\frac{\binom{k_i}{2}\tau_i}{N_e} \right).$$

Note above that we use $\mathrm{N_c}$ *understanding that the times,* τ_i, *are measured in generations. If they are measured in calendar units then* $\mathrm{N_e}$ *would be replaced by* $\mathrm{N_e g_c}$ *where* $\mathrm{g_c}$ *is the generation time in the calendar units employed.*

Furthermore, there is often considerable uncertainty in the reconstructed genealogy. In order to allow for this uncertainty it is necessary to compute the average probability of the population parameters of interest. The calculation involves integrating over genealogies distributed according to the coalescent (Felsenstein 1988, 1992; Griffiths 1989; Griffiths and Tavaré 1994; Kuhner et al. 1995). Integration for some models of interest can be carried out using Monte Carlo methods. Importance-sampling algorithms have been developed to estimate the population parameter $\Theta \propto N_e \mu$ (Griffiths and Tavaré 1994; Stephens and Donnelly 2000), migration rates (Bahlo and Griffiths 2000) and recombination (Fearnhead and Donnelly 2001; Griffiths and Marjoram 1996). Metropolis–Hastings Markov chain Monte Carlo (MCMC) (Hastings 1970; Metropolis et al. 1953) has been used to obtain sample-based estimates of Θ (Kuhner et al. 1995), exponential growth rate (Kuhner et al. 1998), migration rates (Beerli and Felsenstein 1999, 2001) and recombination rate (Kuhner et al. 2000).

In addition to developments in coalescent-based population genetic inference, sequence data sampled at different times are now available from both rapidly evolving viruses such as HIV-1 (Holmes et al. 1993; Rodrigo et al. 1999; Shankarappa et al. 1999; Wolinsky et al. 1996), and from ancient DNA sources (Barnes et al. 2002; Lambert et al. 2002; Leonard et al. 2002; Loreille et al. 2001). Temporally spaced data provide the potential to observe the accumulation of mutations over time, thus allowing the estimation of mutation rate (Drummond and Rodrigo 2000; Rambaut 2000). In fact it is even possible to estimate variation in the mutation rate over time (Drummond et al. 2001). This leads naturally to the more general problem of simultaneous estimation of population parameters and mutation parameters from temporally spaced sequence data (Drummond and Rodrigo 2000; Drummond et al. 2001, 2002; Rodrigo and Felsenstein 1999; Rodrigo et al. 1999).

Non-parametric coalescent methods

A number of non-parametric coalescent methods have been developed to infer population size history from DNA sequences without resorting to simple parametric models of population size history. The main differences among these methods are (1) how the population size function is segmented along the tree, (2) the statistical estimation technique employed and (3) in Bayesian methods, the form of the prior density on the parameters governing the population size function. In the 'classic skyline plot' (Pybus et al. 2000) each coalescent interval is treated as a separate segment, so a tree of n taxa has $n - 1$ population size parameters. However, the true number of population size changes is likely to be substantially fewer, and the generalised skyline plot (Strimmer and Pybus 2001) acknowledges this by grouping the intervals according to the small-sample Akaike information criterion (AICc) (Burnham and Anderson 2002; Hurvich and Tsai 1989). The epidemic history of HIV-2 was investigated using the generalised skyline plot (Strimmer and Pybus 2001), indicating the population size was relatively constant in the early history of HIV-2 subtype A in Guinea-Bissau, before expanding more recently (Lemey et al. 2003). Using this information, the authors then employed a piecewise expansion growth model to estimate the time of expansion to a range of 1955–1970.

While the generalised skyline plot is a good tool for data exploration, and to assist in model selection (for examples see: Lemey et al. 2004; Pybus et al. 2003), it infers demographic history based on a single input tree and therefore does not account for sampling error produced by phylogenetic reconstruction nor for the intrinsic stochasticity of the coalescent process. This shortcoming is overcome by implementing the skyline plot method in a Bayesian statistical framework, which simultaneously infers the sample genealogy, the substitution parameters and the population size history. Further extensions of the generalised skyline plot include modelling the population size by a piecewise-linear function instead of a piecewise-constant population, allowing continuous changes over time rather than sudden jumps. The Bayesian skyline plot (Drummond et al. 2005) has been used to suggest that the effective population size of HIV-1 group M may have grown at a relatively slower rate in the first half of the twentieth century, followed by much faster growth (Worobey et al. 2008). On a much shorter time scale, the Bayesian skyline plot analysis of a data set collected from a pair of HIV-1 donor and recipient was used to reveal a substantial loss of genetic diversity following virus transmission (Edwards et al. 2006). A further parametric analysis assuming constant population size in the donor and logistic growth model in the recipient estimated that more than 99% of the genetic diversity of HIV-1 present in the donor is lost during horizontal transmission. This has important implications as the process underlying the bottleneck determines the viral fitness in the recipient host.

One disadvantage of the Bayesian skyline plot is that the number of changes in the population size has to be specified by the user *a priori* and the appropriate number is seldom known. A solution is provided by methods that perform Bayesian model averaging on the demographic model utilising either reversible-jump MCMC (Opgen-Rhein et al. 2005) or Bayesian variable selection (Heled and Drummond 2008), and in which case the number of population size changes is a random variable estimated as part of the model. Development of non-parametric modelling approaches to the coalescent are ongoing (Gill et al. 2013; Minin et al. 2008; Palacios and Minin 2012, 2013), indicating both the demand for methods to estimate past population sizes and the technical challenge of producing a non-parametric estimator of an inhomogeneous point process that is free of unattractive statistical properties.

The methods for demographic inference discussed so far assume no subdivision within the population of interest. Like changes in the size, population structure can also have an effect on the pattern of the coalescent interval sizes, and thus the reliability of results can be questioned when population structure exists (Pybus et al. 2009). Models of trees with population structure will be discussed in Chapter 5.

2.3.2 Serially sampled coalescent

The serial sample coalescent was introduced by Rodrigo and Felsenstein (1999) and a full Bayesian inference approach to estimating gene trees under the serial sample coalescent was described a few years later (Drummond et al. 2002).

In order to illustrate some of the complexities that occur when introducing non-contemporaneous sequences (i.e. dated tips) we will briefly consider the implications of such sampling to the probability distribution over ranked trees under the constant-size coalescent in a simple example. Recall that for contemporaneous sampling a uniform distribution on ranked trees was induced. For non-contemporaneous sampling this is not the case. Consider the situation in which there are four haploid individuals sampled, two in the present (A, B) and two sampled τ time units in the past, as illustrated in Figure 2.6. The probability that there is no coalescent between A and B closer to the present than τ is:

$$p_{nc} = e^{-\tau/N_e},$$

where N_e is the effective population size. Consequently the probability of coalescence is:

$$p_c = 1 - p_{nc}.$$

If there is a coalescence more recent than time τ then the tree must be one of the following topologies: ((A,B),(C,D)), (((A,B),C),D), (((A,B),D),C).

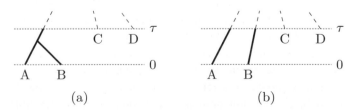

(a)   (b)

Figure 2.6 (a) Only three ranked trees are possible if coalescence of A and B occurs more recently than τ. (b) All ranked topologies are possible if A and B do not coalesce more recently than τ.

Now consider the topology ((A,B),(C,D)). Conditional on coalescence of (A,B) more recent than τ it has a probability of $\frac{1}{3}$. However, if there is no coalescence more recently than τ it has its 'normal' coalescent probability of $\frac{1}{9}$ (being a symmetrical tree shape). This gives a total probability for this tree shape of:

$$P_{((A,B),(C,D))} = \frac{p_c}{3} + \frac{p_{nc}}{9}.$$

Likewise the probability of topologies (((A,B),C),D) and (((A,B),D),C) can be calculated as:

$$P_{(((A,B),C),D)} = P_{(((A,B),D),C)} = \frac{p_c}{3} + \frac{p_{nc}}{18}.$$

The probability of the two remaining symmetrical trees is:

$$P_{((A,C),(B,D))} = P_{((A,D),(B,C))} = \frac{p_{nc}}{9},$$

and the probability of each of the remaining asymmetric trees is $\frac{p_{nc}}{18}$.

Taking $\tau/N_e = 0.5$ then $p_{nc} = 0.607$ and $p_c = 0.393$ gives tree probabilities of:

$$P_{((A,B),(C,D))} \approx 0.199$$
$$P_{(((A,B),C),D)} \approx 0.165$$
$$P_{(((A,B),D),C)} \approx 0.165$$
$$P_{((A,C),(B,D))} \approx 0.0674$$
$$P_{((A,D),(B,C))} \approx 0.0674$$
$$P_{(((C,D),B),A)} \approx 0.0337$$

$$\cdots$$

which is clearly not uniform on ranked topologies. Drummond et al. (2002) describe a method to jointly estimate mutation rate and population size that incorporates the uncertainty in the genealogy of temporally spaced sequences by using MCMC integration.

2.3.3   Modelling epidemic dynamics using coalescent theory

Models that describe epidemic disease progression compartmentalise host individuals into different states. Individuals within each compartment are deemed to be dynamically equivalent. The specific division of the host population depends on the life cycle of the infectious agent in question, spanning a range of scenarios where hosts may or may not be reinfected, possess more than one infection rate, exhibit a period of exposure (incubation period) between becoming infected and becoming infectious, and so forth. Such examples cover the well-known SI (Susceptible-Infected), SIS (Susceptible-Infected-Susceptible), SIR (Susceptible-Infected-Removed) and SEIR (Susceptible-Exposed-Infected-Removed) paradigms (Anderson and May 1991; Keeling and Rohani 2008).

Currently, the probabilities of phylogenetic trees can only be solved analytically for small host population sizes in the simplest endemic setting (SI and SIS) (Leventhal et al. 2014). The SIR model is the simplest that exhibits epidemic dynamics (as opposed to endemic dynamics), by including a third class of 'Removed' individuals who are removed from the infected population by way of acquired immunity, death or some other inability to infect others or to be reinfected. The number of susceptible, infected and removed individuals under an SIR model can be deterministically described forward in time by a trio of coupled ordinary differential equations (ODEs), where β and μ respectively represent the transition rates from susceptible S to infected I, and infected I to removed R, such that

$$\frac{\mathrm{d}}{\mathrm{d}\tau}S(\tau) = -\beta I(\tau)S(\tau), \tag{2.6}$$

$$\frac{\mathrm{d}}{\mathrm{d}\tau}I(\tau) = \beta I(\tau)S(\tau) - \mu I(\tau), \tag{2.7}$$

$$\frac{\mathrm{d}}{\mathrm{d}\tau}R(\tau) = \mu I(\tau). \tag{2.8}$$

Considering this model in the reverse time direction, we have $\mathcal{S}(t) = S(z_0 - t)$, $\mathcal{I}(t) = I(z_0 - t)$, $\mathcal{R}(t) = R(z_0 - t)$, from an origin z_0 ago.

Recall that the coalescent calculates the probability density of a tree given the coalescent rate. The coalescent rate for k lineages is $\binom{k}{2}$ times the inverse of the product of effective population size N_e and generation time g_c. Volz (2012) proposed a coalescent approximation to epidemiological models such as the SIR, where the effective population size N_e is the expected number of infected individuals through time, and the generation time g_c is derived as follows.

We summarise the epidemiological parameters $\eta = \{\beta, \mu, S_0, z_0\}$ with the start of the epidemic being at time z_0 in the past and S_0 being the susceptible population size at time z_0. The expected epidemic trajectory obtained from Equations (2.6)–(2.8) is denoted $\mathcal{V} = (\mathcal{S}, \mathcal{I}, \mathcal{R})$. Let $f(g|\mathcal{V}, \eta)$ be the probability density of a tree given the expected trajectory and the epidemic parameters η. The coalescent rate $\lambda_k(t)$ of k co-existing lineages computed following Volz (2012) is:

$\lambda_k(t) =$ (Prob. of coalescence given single birth in population) $\times$ (total birth rate)

$$= \binom{k}{2}\left(\frac{\mathcal{I}(t)}{2}\right)^{-1} \times \beta\mathcal{S}(t)\mathcal{I}(t)$$

$$\approx \binom{k}{2}\frac{2\beta\mathcal{S}(t)}{\mathcal{I}(t)}. \tag{2.9}$$

Note that 'birth' in this context refers to an increase in the number of infected hosts by the infection process. In the last line, the approximation $\mathcal{I}(t) \approx \mathcal{I}(t) - 1$ is used, following (Volz 2012).

In summary, the relationship between the estimators of $N_e g_c(t)$, such as the Bayesian skyline plot, and the coalescent model approximating the SIR dynamics described by Volz (2012), is:

$$\binom{k}{2}\frac{1}{N_e(t)g_c(t)} = \lambda_k(t) \approx \binom{k}{2}\frac{2\beta\mathcal{S}(t)}{\mathcal{I}(t)}, \tag{2.10}$$

where $N_e(t) = \mathcal{I}(t)$ is a deterministically varying population size, and $g_c(t) \approx \frac{1}{2\beta S(t)}$ is a deterministically varying generation time that results from the slowdown in infection rate per lineage as the susceptible pool is used up.

The probability density of coalescent intervals **t** given an epidemic description can then be easily computed, following Equation (2.5) and using calendar time units:

$$f(\mathbf{t}|\mathcal{V}, \eta) = \prod_{i \in Y} \lambda_{k_i}(t_i) \prod_{i \in V} \exp\left(-\int_{t_{i-1}}^{t_i} \lambda_{k_i}(x)\mathrm{d}x\right). \qquad (2.11)$$

2.4   Birth–death models

The coalescent can be derived by considering the data set to be a small sample from an idealised Wright–Fisher population. Coalescent times only depend on deterministic population size, meaning population size is the only parameter in the coalescent.

If the assumption of small sample size or deterministic population size breaks down (Popinga et al. 2015; Stadler et al. 2015) or parameters other than population size may governthe distribution of time-trees, then different models need to be considered. Classically, forward in time birth–death processes (Kendall 1948) are used as alternatives to the coalescent. We will first discuss birth–death processes with constant rates and samples taken at one point in time, which includes the Yule model (Harding 1971; Yule 1924). This model is then generalised to allow for changing rates, and finally sequential sampling is considered.

2.4.1   Constant-rate birth–death models

The continuous-time constant-rate birth–death process is a birth–death process which starts with one lineage at time z_0 in the past and continues forward in time with a stochastic rate of birth (λ) and a stochastic rate of death (μ) until the present (time 0). At present, each extant lineage is sampled with probability ρ. The process generates a tree with extinct and extant lineages which we denote as 'complete tree'. Typically the extinct and the non-sampled lineages are deleted, producing a 'reconstructed tree' on only sampled extant lineages (see Figure 2.7).

This reconstructed tree, a time-tree g produced after time z_0, may have any number n of tips ($n = 0, 1, 2, \ldots$), i.e. $g \in \mathcal{G}$ where $\mathcal{G} = \mathcal{G}_0 \cup \mathcal{G}_1 \cup \mathcal{G}_2 \ldots$. In order to investigate a particular time-tree under the constant-rate birth–death model we need to compute the probability density of the time-tree $g \in \mathcal{G}$ given a birth rate λ, a death rate μ, a sampling probability ρ and a time of origin z_0 (Stadler 2010):

$$f_{BD}(g|\lambda, \mu, \rho, z_0) = p_1(z_0) \prod_{i=1}^{n-1} \lambda p_1(t_{n+i})$$

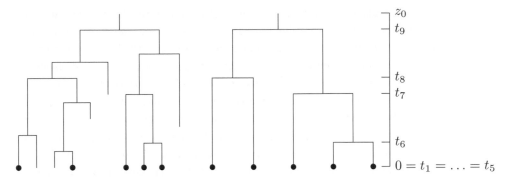

Figure 2.7 Complete tree of age z_0 (left) and corresponding reconstructed tree (right). In the reconstructed phylogeny, all extinct and non-sampled extant species are pruned, only the sampled species (denoted with a black circle) are included.

with

$$p_1(x) = \frac{\rho(\lambda - \mu)^2 e^{-(\lambda - \mu)x}}{(\rho\lambda + (\lambda(1 - \rho) - \mu)e^{-(\lambda - \mu)x})^2},$$

where $p_1(x)$ is the probability of an individual at time x in the past having precisely one sampled descendant.

When assuming additionally a probability density $f(z_0)$ for z_0, we obtain the probability density of the time-tree $g \in \mathcal{G}$ together with its time of origin z_0, given a birth rate λ, a death rate μ and a sampling probability ρ:

$$f_{BD}(g, z_0 | \lambda, \mu, \rho) = f_{BD}(g | \lambda, \mu, \rho, z_0)f(z_0).$$

We made the assumption here that the distribution of z_0 is independent of λ, μ, ρ, i.e. $f(z_0 | \lambda, \mu, \rho) \equiv f(z_0)$, following the usual assumption of independence of hyper-prior distributions.

For obtaining non-biased estimates, we suggest to condition the process on yielding at least one sample (i.e. $g \in \mathcal{G}_S$ with $\mathcal{G}_S = \mathcal{G}_1 \cup \mathcal{G}_2 \cup \mathcal{G}_3 \ldots$), as we only consider data sets which contain at least one sample (Stadler 2013b). We obtain such a conditioning through dividing $f_{BD}(g, z_0 | \lambda, \mu, \rho)$ by the probability of obtaining at least one sample. We define $p_0(x)$ to be the probability of an individual at time x in the past not having any sampled descendants, and thus $1 - p_0(x)$ is the probability of at least one sample (Yang and Rannala 1997):

$$1 - p_0(x) = \frac{\rho(\lambda - \mu)}{\rho\lambda + (\lambda(1 - \rho) - \mu)e^{-(\lambda - \mu)x}}.$$

The probability density of the time-tree $g \in \mathcal{G}$ together with its time of origin z_0, given a birth rate λ, a death rate μ, a sampling probability ρ and conditioned on at least one sample (S) is:

$$f_{BD}(g, z_0 | \lambda, \mu, \rho; S) = f(z_0) \frac{p_1(z_0)}{1 - p_0(z_0)} \prod_{i=1}^{n-1} \lambda p_1(t_{n+i}). \qquad (2.12)$$

Equation (2.12) defines the probability density of the time-tree $g \in \mathcal{G}_S$. Thus when performing parameter inference, the number of samples n is considered as part of the data. In contrast, under the coalescent, the probability density of the time-tree $g \in \mathcal{G}_n$ is calculated (Equation (2.5)). Thus, parameter inference is conditioned on the number of samples n, which means that we do not use the information n for inference and thus ignore some information in the data by conditioning on it.

In order to compare the constant-rate birth–death process to the coalescent, we may want to condition on observing n tips. Such trees correspond to simulations where first a time z_0 is sampled, and then a tree with age z_0 is being simulated but only kept if the final number of tips is n. The probability density of the time-tree $g \in \mathcal{G}_n$ together with its time of origin z_0, given λ, μ and ρ, is (Stadler 2013b):

$$f_{BD}(g, z_0 | \lambda, \mu, \rho; n) = f_{BD}(g | \lambda, \mu, \rho, z_0; n) f(z_0; n), \qquad (2.13)$$

with

$$f_{BD}(g | \lambda, \mu, \rho, z_0; n) = \prod_{i=1}^{n-1} \lambda \frac{p_1(t_i)}{q(z_0)},$$

where

$$q(z_0) = \rho \lambda (1 - e^{-(\lambda - \mu) z_0}) / (\lambda \rho + (\lambda (1 - \rho) - \mu) e^{-(\lambda - \mu) z_0}).$$

Note that when we condition on n (instead of considering it as part of the data), the distribution for z_0 as well as all other hyper-prior distributions are formally chosen with knowledge of n.

We conclude the section on the constant-rate birth–death model with several remarks: The constant-rate birth–death model with $\mu = 0$ and $\rho = 1$, i.e. no extinction and complete sampling, corresponds to the well-known Yule model (Edwards 1970; Yule 1924).

The three parameters λ, μ, ρ are non-identifiable, meaning that the probability density of a time-tree is determined by the two parameters $\lambda - \mu$ and $\lambda \rho$ (the probability density only depends on these two parameters if the probability density is conditioned on survival or on n samples (Stadler 2009)). Thus, if the priors for all three parameters are non-informative, we obtain large credible intervals when estimating these parameters.

One may speculate that the distribution of $g \in \mathcal{G}_n$ under the constant-rate birth–death process (Equation (2.13)) for $\rho \to 0$ converges to the distribution of $g \in \mathcal{G}_n$ under the coalescent (Equation (2.5)). However, this is only true in expectation under special assumptions (Gernhard 2008; Stadler 2009). One difference between the constant-rate birth–death process limit and the coalescent is that the birth–death process induces a stochastically varying population size while classic coalescent theory relies on a deterministic population size. Recent work demonstrates that even when coalescent theory is applied to a population with a stochastically varying population size, the result is not equivalent to the constant-rate birth–death limit (Stadler et al. 2015).

2.4.2   Time-dependent birth–death models

Recently, the assumptions of the constant-rate birth–death model have been relaxed, allowing for birth and death rates to change through time, which is an analogue to the coalescent with a population size changing through time (see Equation (2.5)). Morlon et al. (2011, Equation 1) derived a general expression for the probability density of the time-tree $g \in \mathcal{G}$ when $\lambda(x)$ and $\mu(x)$ are birth and death rates changing as a function of time x, $f_{BD}(g|\lambda(x), \mu(x), \rho, z_0; S)$. Note that an equivalent expression for trees starting with the first split (rather than at time z_0) has been derived by Nee et al. (1994, Equation 20). For general $\lambda(x)$ and $\mu(x)$, the equation for $f_{BD}(g|\lambda(x), \mu(x), \rho, z_0; S)$ contains integrals over time which need to be evaluated numerically. The time-dependent macro-evolutionary model allows, for example, to reconcile the molecular phylogeny of cetaceans (whales, dolphins and porpoises) with the fossil record (Morlon et al. 2011).

The integrals can be solved analytically, assuming piecewise constant rates which change at time points $x_- = (x_1, x_2, \ldots, x_m)$ where $x_i > x_{i+1}$, $x_0 := z_0$ and $x_{m+1} := 0$. The birth (resp. death) rate in (x_i, x_{i+1}) is λ_i (resp. μ_i). We write $\lambda_- = (\lambda_0, \lambda_1, \ldots, \lambda_m)$ and $\mu_- = (\mu_0, \mu_1, \ldots, \mu_m)$. Theorem 2.6 in (Stadler 2011) provides an analytic expression for $f_{BD}(g|\lambda_-, \mu_-, \rho, x_-, z_0)$, which needs to be divided by $(1 - p_0(z_0))$ (with $p_0(z_0)$ being provided in (Stadler 2011, Equation 1)), in order to obtain $f_{BD}(g|\lambda_-, \mu_-, \rho, x_-, z_0; S)$. This piecewise constant birth–death process is analogue to the coalescent skyline plot where the population size is piecewise constant, thus we call this model the birth–death skyline plot (Stadler et al. 2013). The birth–death skyline plot has been used, for example, to reject the hypothesis of increased mammalian diversification following the K/T boundary (Meredith et al. 2011; Stadler 2011).

General birth–death models

So far we have considered constant or time-dependent birth and death rates. Further extensions of the constant-rate birth–death model have been developed, for example assuming that the birth and death rates are dependent on the number of existing lineages (diversity-dependent diversification, Etienne et al. 2012; Rabosky 2007) or assuming that the birth and death rates are dependent on the trait of a lineage (trait-dependent diversification, FitzJohn 2010; FitzJohn et al. 2009; Jones 2011; Maddison 2007)).

Aldous (2001) showed that for all birth–death processes under which individuals are exchangeable (meaning if a birth, death or sampling event happens, each species is equally likely to be the one undergoing this event), then for a vector $t_{n+1}, \ldots, t_{2n-1}$ of branching times together with tree age z_0, each ranked tree is equally likely, meaning we can consider the discrete ranked tree independent of the continuous branching time vector. Stadler (2013a) generalised this result to processes which are species-speciation-exchangeable. Recall that the same conclusion was drawn under the coalescent, and that the ranked tree together with the branching times define the time-tree. Out of all birth–death models mentioned above only the trait-dependent model does not satisfy the species-speciation-exchangeable assumption.

2.4.3   Serially sampled birth–death models

Since in the previous section we only defined a sampling probability ρ at present, the birth–death model only gives rise to time-trees with contemporaneous tips. Stadler (2010) extended the constant rate birth–death model to account for serial sampling by assuming a sampling rate ψ. This means that each lineage is sampled with a rate ψ and it is assumed that the lineage goes extinct after being sampled.

For constant λ, μ, ρ, ψ the probability density of the time-tree $g \in \mathcal{G}$,

$$f_{BD}(g|\lambda, \mu, \rho, \psi, z_0; S),$$

is given in (Stadler 2010, Corollary 3.7). This model was used to quantify the basic reproductive number for HIV in Switzerland (Stadler et al. 2012).

We note that if $\mu = 0$, we obtain time-trees with all extinct lineages being included. If $\rho = 0$, then it is straightforward to show that the probability density of a time-tree only depends on two parameters $\lambda - \mu - \psi$ and $\lambda\psi$, if conditioned on survival (Stadler 2010).

Again, the model can be extended to piecewise changing rates (birth–death skyline plot), and the probability density of the time-tree $g \in \mathcal{G}$,

$$f_{BD}(g|\lambda_-, \mu_-, \rho_-, \psi_-, x_-, z_0; S),$$

is given in (Stadler et al. 2013, Theorem 1). The birth–death skyline plot was used to recover epidemiological dynamics of HIV in the UK and HCV in Egypt.

Recall that when analysing serially sampled data using the coalescent, we condition the analysis on the number of samples n as well as on the sampling times. The birth–death model, however, treats n and the sampling times as part of the data, and thus the parameters are informed by the sampling times of the particular data sets.

Recent work started incorporating diversity-dependent models (Kühnert et al. 2014; Leventhal et al. 2014) and trait-dependent models (Stadler and Bonhoeffer 2013) for serially sampled data. Diversity-dependent models can be used to explicitly model epidemiological dynamics in infectious diseases by acknowledging the dependence of transmission rates on the number of susceptible individuals, formalised in SI, SIS, SIR or SEIR models described in Section 2.3.3. In such models the birth rate of the birth–death model is $\lambda = \beta S$, with β being the transition rate from susceptible to infected and S being the number of susceptibles. Trait-dependent models may be used for structured populations, where different population groups are characterized by a trait. Chapter 5 discusses some more details of these phylodynamic models.

2.5   Trees within trees

2.5.1   The multispecies coalescent

So far we assumed that the genealogy equals the species or transmission tree. However, this is an approximation. The genealogy is actually embedded within the species/ transmission tree.

Table 2.3 Notation for multispecies coalescent

Notation	Description
n_S	The number of species in a species tree.
t_i	The speciation time i in the species tree.
g_S	The time-tree representing the species tree topology and speciation times.
$\mathbf{N}$	$= \{N_1(t), N_2(t), \ldots, N_{2n_S-1}(t)\}$, a set of population size functions. $N_i(t)$ is the population size at time t in the ith branch of the species tree.
$S = \langle g_S, \mathbf{N} \rangle$	The species (or population) tree, made up of a time-tree g_S and a set of population size functions, $\mathbf{N}$ (one population size function for each branch in the time-tree, including the root branch).
$L_{i,S}(g)$	The set of gene tree coalescent intervals for genealogy g that are contained in the ith branch of species tree S.

The multispecies coalescent brings together coalescent and birth–death models of time-trees into a single model. It describes the probability distribution of one or more gene trees that are nested inside a *species tree*. The species tree describes the relationship between the sampled species or, sometimes, sampled populations that have been separated for long periods of time relative to their population sizes. In the latter case it may be referred to as a *population tree* instead. The multispecies coalescent model can be used to estimate the species time-tree g_S, together with ancestral population sizes $\mathbf{N}$, given the sequence data from multiple genes, whose gene trees may differ due to incomplete lineage sorting (Pamilo and Nei 1988).

The probability of a gene tree with respect to a species tree S, using notation in Table 2.3, is:

$$f(g|S) = \prod_{i=1}^{2n_S-1} f(L_{i,S}(g)|N_i(t)). \qquad (2.14)$$

Figure 2.8 shows a gene tree of $n = 10$ taxa samples from $n_S = 3$ species.

The joint probability distribution of the species tree and the gene tree can be written:

$$\begin{aligned} f(g, S) &= f(g|S)f(S) \\ &= f(g|g_S, \mathbf{N})f(g_S)f(\mathbf{N}), \end{aligned} \qquad (2.15)$$

where $f(g_S)$ is typically a Yule or birth–death prior on the species time-tree and $f(\mathbf{N})$ is a prior on the population size functions. For information on prior formulations for population sizes and Bayesian inference under this model, see Section 8.3.

2.5.2   Viral transmission histories

Another context in which a hierarchy of nested time-trees can be estimated from a single data set is in the case of estimating transmission histories from viral sequence data, where the nested time-tree is the viral gene tree, and the encompassing tree is the tree describing transmissions between hosts (Leitner and Fitch 1999). Didelot et al. (2014) and Romero-Severson et al. (2014) point out that the common assumption of the gene

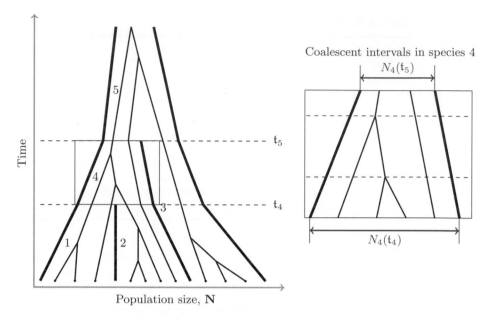

Figure 2.8 A species tree on $n_S = 3$ species with a gene tree of $\mathbf{n} = (3, 4, 3)$ samples embedded.

tree being equal to the transmission history is problematic and may bias transmission time estimates.

In this interpretation, we argue that the natural generative model at the level of the host population is a branching process of infections, where each branching event represents a transmission of the disease from one infected individual to the next, and the terminal branches of this transmission tree represent the transition from infectious to recovery or death of the infected host organism. For multicellular host species there is an additional process of proliferation of infected cells within the host's body (often restricted to certain susceptible tissues) that also has a *within-host* branching process of cell-to-cell infections. This two-level hierarchical process can be extended to consider different infectious compartments at the host level, representing different stages in disease progression, and/or different classes of dynamic behaviour among hosts.

Accepting the above as the basic schema for the generative process, one needs to also consider a typical observation process of an epidemic or endemic disease. It is often the case that data are obtained through time from some fraction, but not all, of the infected individuals. Figure 2.9 illustrates the relationship between the full transmission history and the various sampled histories. The transmission tree prior may be one of the birth–death models introduced in Section 2.4. The gene tree prior may be a coalescent process within the transmission tree (analogous to a coalescent process within a species tree). Notice that the sampled host transmission tree may have internal nodes with a single child lineage representing a direct ancestor of a subsequent sample. We refer to such trees as *sampled ancestor trees* (Gavryushkina et al. 2013) and a reversible-jump Bayesian inference scheme for such trees has recently been described (Gavryushkina et al. 2014).

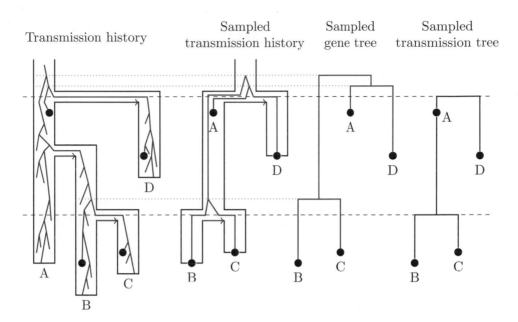

Transmission history   Sampled transmission history   Sampled gene tree   Sampled transmission tree

Figure 2.9 (1) A transmission tree and embedded pathogen gene tree; (2) the sampled transmission tree; (3) the sampled pathogen gene tree; (4) the sampled host transmission tree.

This model schema (combining both the generative and observational processes) can be readily simulated with a recently developed BEAST 2 package called MASTER (Vaughan and Drummond 2013), and inference approaches under models of similar form have been described (Ypma et al. 2013), but full likelihood inference under the model depicted in Figure 2.9 is not yet available.

2.6   Exercise

Tajima (1983) described the probability of different *evolutionary relationships* on a sample of n genes. His trees were unlabelled but with ranked internal nodes and thus could be called *ranked tree shapes*. With increasing n, the number of ranked tree shapes grows faster than (unranked) tree shapes and slower than ranked labelled trees. Let u_n be the number of ranked tree shapes for n taxa. It is easy to see that $u_2 = u_3 = 1$. Tajima Tajima (1983) enumerated u_4 and u_5 but did not give a general formula for counting the number of ranked tree shapes for arbitrary size n. Below is a partially completed sequence of values up to u_{10}:

$$(u_2, u_3, \ldots, u_{10}) = (1, 1, 2, 5, ?, 61, 272, 1385, ?).$$

Can you fill in one or both of the missing values in the sequence? Show your working.

3   Substitution and site models

The simplest measure of distance between a pair of aligned molecular sequences is the number of sites at which they differ. This is known as the Hamming distance (h). This raw score can be normalised for the length of a sequence (l) to get the proportion of sites that differ between the two sequences, $p = h/l$. Consider two hypothetical nucleotide fragments of length $l = 10$:

Sequence 1	A A T C T G T G T G
Sequence 2	A G C C T G G G T A

In these sequences $h = 4$ and $p = 4/10 = 0.4$. The proportion of sites that are different, p, is an estimate of the evolutionary distance between these two sequences. A single nucleotide site can, given enough time, undergo multiple substitution events. Because the alphabet of nucleotide sequences is small, multiple substitutions can be hidden by reversals and parallel changes. If this is the case, some substitutions will not be observed. Therefore the estimate of 0.4 substitutions/site in this example could be an underestimate. This is easily recognised if one considers two hypothetical sequences separated by a very large evolutionary distance – for example ten substitutions/site. Even though the two sequences will be essentially random with respect to each other, they will still, by chance alone, have matches at about 25% of the sites. This would give them an uncorrected distance, p, of 0.75 substitutions/site, despite being actually separated by ten substitutions/site.

To compensate for this tendency to underestimate large evolutionary distances, a technique called *distance correction* is used. Distance correction requires an explicit model of molecular evolution. The simplest of these models is the Jukes–Cantor (JC) model (Jukes and Cantor 1969). Under the JC model, an estimate for the evolutionary distance between two nucleotide sequences is:

$$\hat{d} = -\frac{3}{4} \ln \left(1 - \frac{4}{3}p \right).$$

For the example above, the estimated genetic distance $\hat{d} \approx 0.571605$, and this is an estimate of the expected number of substitutions per site. This model assumes that all substitutions are equally likely and that the frequencies of all nucleotides are equal and at equilibrium. This chapter describes the JC model and related continuous-time

Markov processes (CTMPs). In general these models are assumed to act independently and identically across sites in a sequence alignment.

3.1 Continuous-time Markov process

A CTMP is a stochastic process taking values from a discrete state space at random times, and which satisfies the Markov property.

Let $X(t)$ be the random variable representing the state of a Markov process at time t. Assuming that the Markov process is in state $i \in \{A, C, G, T\}$ at time t, then in the next small moment, the probability that the process transitions to state $j \in \{A, C, G, T\}$ is governed by the *instantaneous transition rate matrix Q*:

$$
Q = \begin{bmatrix}
\cdot & q_{AC} & q_{AG} & q_{AT} \\
q_{CA} & \cdot & q_{CG} & q_{CT} \\
q_{GA} & q_{GC} & \cdot & q_{GT} \\
q_{TA} & q_{TC} & q_{TG} & \cdot
\end{bmatrix},
$$

with the diagonal entries $q_{ii} = -\sum_{j \neq i} q_{ij}$, so that the rates for a given state should sum to zero across the row. The off-diagonal entries $q_{ij} > 0, i \neq j$ are positive and represent rates of flow from nucleotide i to nucleotide j. The diagonal entries q_{ii} represent the total flow out of nucleotide state i and are thus negative. At equilibrium, the total rate of change per site per unit time is thus:

$$
\mu = -\sum_i \pi_i q_{ii},
$$

where π_i is the probability of being in state i at equilibrium, so that μ is just the weighted average outflow rate at equilibrium.

Figure 3.1 depicts the states and instantaneous transition rates in a general Markov model on the DNA alphabet. In particular, for a small time Δt we have:

$$
\Pr\{X(t + \Delta t) = j | X(t) = i\} = q_{ij} \Delta t.
$$

In general, the *transition probability matrix*, $P(t)$, provides the probability for being state j after time t, assuming the process began in state i:

$$
P(t) = \exp(Qt).
$$

For simple models, such as the Jukes–Cantor model (see following section) the elements of the transition probability matrix have analytical closed-form solutions. However, for more complex models (including general time-reversible; GTR) this matrix exponentiation can only be computed numerically, typically by Eigen decomposition (Stewart 1994).

3.1.1 Time-reversible and stationary

For the purposes of modelling substitution processes, time-reversible and stationary Markov processes are mostly used. Both of these properties are valuable principally

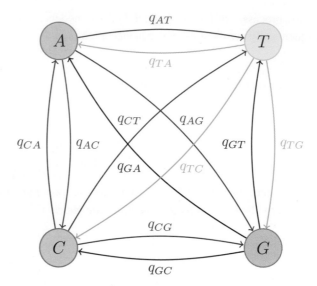

Figure 3.1 Transition rates for a DNA model of evolution.

because they describe processes that are more mathematically tractable, as opposed to biologically realistic.

A stationary CTMP has the following properties:

$$\pi Q = 0, \quad \pi P(t) = \pi, \quad \forall t,$$

where $\pi = [\pi_A, \pi_C, \pi_G, \pi_T]$ is the equilibrium distribution of base frequencies. In addition, a time-reversible CTMP satisfies the detailed balance property:

$$\pi_i P(t)_{ij} = \pi_j P(t)_{ji}, \quad \forall i, j, t.$$

3.2   DNA models

DNA substitution models are specified over a four-letter alphabet $\mathsf{C} = \{A, C, G, T\}$ with 12 transition rates in the instantaneous rate matrix (see Figure 3.1). This section describes the construction of the rate matrix for a number of the common named DNA substitution models as well as calculation of the resulting transition probabilities where a closed-form solution exists.

3.2.1   Jukes–Cantor

The Jukes–Cantor process (Jukes and Cantor 1969) is the simplest CTMP. All transitions have equal rates, and all bases have equal frequencies ($\pi_A = \pi_C = \pi_G = \pi_T = 1/4$). An unnormalised $\hat{Q}$ matrix for the Jukes–Cantor model is:

$$\hat{Q} = \begin{bmatrix} -3 & 1 & 1 & 1 \\ 1 & -3 & 1 & 1 \\ 1 & 1 & -3 & 1 \\ 1 & 1 & 1 & -3 \end{bmatrix}.$$

However, it is customary when describing substitution processes with a CTMP to used a normalised instantaneous rate matrix $Q = \beta \hat{Q}$, so that the normalised matrix has an expected mutation rate of 1 per unit time, i.e. $\mu = -\sum_i \pi_i q_{ii} = 1$. This can be achieved by choosing $\beta = 1/ -\sum_i \pi_i \hat{q}_{ii}$. For the above $\hat{Q}$ matrix, setting $Q = \frac{1}{3}\hat{Q}$ leads to a normalised Q matrix for the Jukes–Cantor model of:

$$Q = \begin{bmatrix} -1 & 1/3 & 1/3 & 1/3 \\ 1/3 & -1 & 1/3 & 1/3 \\ 1/3 & 1/3 & -1 & 1/3 \\ 1/3 & 1/3 & 1/3 & -1 \end{bmatrix}.$$

Notice that this matrix has no free parameters. The benefit of normalising to unitary output is that the times to calculate transition probabilities for can now be expressed in substitutions per site (i.e. genetic distances). The entries of the transition probability matrix for the Jukes–Cantor process are easily computed and as follows:

$$p_{ij}(d) = \begin{cases} \frac{1}{4} + \frac{3}{4}\exp\left(-\frac{4}{3}d\right) & \text{if } i = j \\ \frac{1}{4} - \frac{1}{4}\exp\left(-\frac{4}{3}d\right) & \text{if } i \neq j \end{cases}, \tag{3.1}$$

where d is the evolutionary time in units of substitutions per site. The transition probabilities from nucleotide A are plotted against genetic distance (d) in Figure 3.2. It can be seen that at large genetic distances the transition probabilities all asymptote to $\frac{1}{4}$, reflecting the fact that for great enough evolutionary time, all nucleotides are

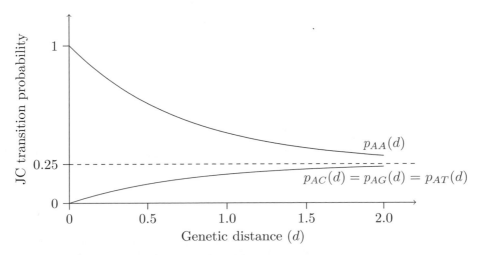

Figure 3.2 The transition probabilities from nucleotide A for the Jukes–Cantor model, plotted against genetic distance ($d = \mu t$).

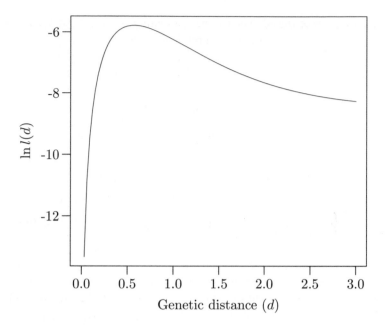

Figure 3.3 The log likelihood curve for genetic distance between two sequences of length 10 and differing at four sites.

equally probable, regardless of the initial nucleotide state. Armed with the transition probabilities in Equation (3.1) it is possible to develop a probability distribution over Hamming distance h for a given genetic distance (d) between two sequences of length l:

$$L(d) = \Pr(h|d) = \binom{l}{h} p_{ii}(d)^{(l-h)} \left[1 - p_{ii}(d)\right]^h .$$

Using the example from the beginning of the chapter, where the two sequences were of length $L = 10$ and differed at $H = 4$ sites the (log-)likelihood as a function of d is shown in Figure 3.3.

3.2.2   K80

The K80 model (Kimura 1980) distinguishes between *transitions* ($A \longleftrightarrow G$ and $C \longleftrightarrow T$ state changes) and *transversions* (state changes from a purine to pyrimidine or vice versa). The model assumes base frequencies are equal for all characters. This transition/transversion bias is governed by the κ parameter and the Q matrix is:

$$Q = \beta \begin{bmatrix} -2-\kappa & 1 & \kappa & 1 \\ 1 & -2-\kappa & 1 & \kappa \\ \kappa & 1 & -2-\kappa & 1 \\ 1 & \kappa & 1 & -2-\kappa \end{bmatrix}.$$

The normalised Q is obtained by setting $\beta = \frac{1}{2+\kappa}$. Note that this model has one free parameter, κ. The transition probabilities are:

$$p_{ij}(d) = \begin{cases} \frac{1}{4} + \frac{1}{4}\exp(-\frac{4}{\kappa+2}d) + \frac{1}{2}\exp(-\frac{2\kappa+2}{\kappa+2}d) & \text{if } i=j \\ \frac{1}{4} + \frac{1}{4}\exp(-\frac{4}{\kappa+2}d) - \frac{1}{2}\exp(-\frac{2\kappa+2}{\kappa+2}d) & \text{if transition} \\ \frac{1}{4} - \frac{1}{4}\exp(-\frac{4}{\kappa+2}d) & \text{if transversion} \end{cases} .$$

Figure 3.4 plots the first row in the K80 transition probability matrix as a function of genetic distance, for $\kappa = 4$.

3.2.3   F81

The F81 model (Felsenstein 1981) allows for unequal base frequencies ($\pi_A \neq \pi_C \neq \pi_G \neq \pi_T$). The Q matrix is:

$$Q = \beta \begin{bmatrix} \pi_A - 1 & \pi_C & \pi_G & \pi_T \\ \pi_A & \pi_C - 1 & \pi_G & \pi_T \\ \pi_A & \pi_C & \pi_G - 1 & \pi_T \\ \pi_A & \pi_C & \pi_G & \pi_T - 1 \end{bmatrix} .$$

The normalised Q is obtained by setting $\beta = \frac{1}{1-\sum_i \pi_i^2}$. Note that this model has three free parameters, since the constraint that the four equilibrium frequencies must sum to 1 ($\pi_A + \pi_C + \pi_G + \pi_T = 1$) removes one degree of freedom. The transition probabilities are:

$$p_{ij}(d) = \begin{cases} \pi_i + (1 - \pi_i)e^{-\beta d} & \text{if } i=j \\ \pi_j(1 - e^{-\beta d}) & \text{if } i \neq j \end{cases} .$$

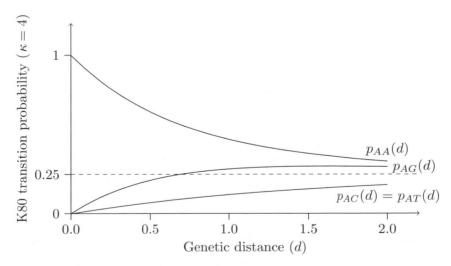

Figure 3.4 The transition probabilities from nucleotide A for the K80 model with $\kappa = 4$, plotted against genetic distance ($d = \mu t$). Although, in this case, $p_{AG}(d)$ exceeds $\frac{1}{4}$ at genetic distances above $d = \ln 2 \approx 0.693147$, all transition probabilities still asymptote to $\frac{1}{4}$ for very large d.

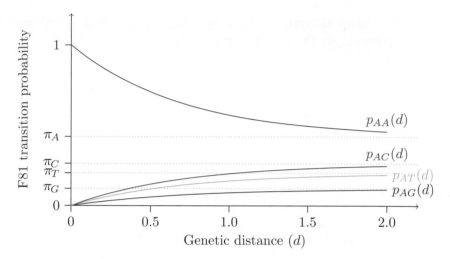

Figure 3.5 The transition probabilities from nucleotide A for the F81 model with $\pi = (\pi_A = 0.429, \pi_C = 0.262, \pi_G = 0.106, \pi_T = 0.203)$, plotted against genetic distance ($d = \mu t$).

Figure 3.5 shows how these transition probabilities asymptotically approach the equilibrium base frequency of the final state with increasing genetic distance.

3.2.4 HKY

The HKY process (Hasegawa et al. 1985) was introduced to better model the substitution process in primate mtDNA. The model combines the parameters in the K80 and F81 models to allow for both unequal base frequencies and a transition/transversion bias. The Q matrix has the following structure:

$$Q = \beta \begin{bmatrix} \cdot & \pi_C & \kappa\pi_G & \pi_T \\ \pi_A & \cdot & \pi_G & \kappa\pi_T \\ \kappa\pi_A & \pi_C & \cdot & \pi_T \\ \pi_A & \kappa\pi_C & \pi_G & \cdot \end{bmatrix}.$$

The diagonal entries are omitted for clarity, but as usual are set so that the rows sum to zero. The transition probabilities for the HKY model can be computed and have a closed-form solution; however, the formulae are rather long-winded and are omitted for brevity.

3.2.5 GTR

The GTR model is the most general time-reversible stationary CTMP for describing the substitution process. The Q matrix is:

$$Q = \beta \begin{bmatrix} \cdot & a\pi_C & b\pi_G & c\pi_T \\ a\pi_A & \cdot & d\pi_G & e\pi_T \\ b\pi_A & d\pi_C & \cdot & \pi_T \\ c\pi_A & e\pi_C & \pi_G & \cdot \end{bmatrix}.$$

The normalised Q is obtained by setting $\beta = 1/[2(a\pi_A\pi_C + b\pi_A\pi_G + c\pi_A\pi_T + d\pi_C\pi_G + e\pi_C\pi_T + \pi_G\pi_T)]$. The transition probabilities do not have a simple closed-form solution and thus a numerical approach to the matrix exponentiation is required.

3.3   Codon models

Protein-coding genes have a natural pattern due to the genetic code that can be exploited by extending a four-nucleotide state space to a 64-codon state space.

Two pairs of researchers published papers on codon-based Markov models of substitution in the same volume in 1994 (Goldman and Yang 1994; Muse and Gaut 1994). The key features they shared were:

- a 61-codon state space (excluding the three stop codons);
- a zero rate for substitutions that changed more than one nucleotide in a codon at any given instant (so each codon has nine immediate neighbours, minus any stop codons);
- a synonymous/nonsynonymous bias parameter making synonymous mutations, that is mutations that do not change the protein that the codon codes for, more likely than nonsynonymous mutations.

Given two codons $i = (i_1, i_2, i_3)$ and $j = (j_1, j_2, j_3)$ the Muse–Gaut-94 codon model has the following entries in the Q matrix:

$$q_{ij} = \begin{cases} \beta\omega\pi_{j_k} & \text{if nonsynonymous change at codon position } k \\ \beta\pi_{j_k} & \text{if synonymous change at codon position } k \\ 0 & \text{if codons } i \text{ and } j \text{ differ at more than one position} \end{cases}.$$

In addition the Goldman–Yang-94 model includes a transition/transversion bias giving rise to entries in Q as follows:

$$q_{ij} = \begin{cases} \beta\kappa\omega\pi_j & \text{if nonsynonymous transition} \\ \beta\omega\pi_j & \text{if nonsynonymous transversion} \\ \beta\kappa\pi_j & \text{if synonymous transition} \\ \beta\pi_j & \text{if synonymous transversion} \\ 0 & \text{if codons differ at more than one position} \end{cases}.$$

Notice that in the Goldman–Yang model the equilibrium distribution is parameterised in terms of 61 codon frequencies, whereas in the Muse–Gaut model the equilibrium distribution is parameterised in terms of four nucleotide frequencies.

Finally, in both of these models, if β is chosen so that Q has unitary output then branches will be in units of substitutions *per codon* rather than per site. So, in order to have branches of the same scale as in nucleotide models, a codon-based Q matrix should be scaled to have a total rate of three per unit time, instead of one, i.e. $-\sum_i \pi_i q_{ii} = 3$.

3.4   Microsatellite models

A microsatellite (or short tandem repeat; STR) is a region of DNA in which a short DNA sequence motif (length one to six nucleotides) is repeated in an array, e.g. the sequence AGAGAGAGAGAGAG is a dinucleotide microsatellite comprising seven repeats of the motif AG. Because they are abundant, widely distributed in the genome and highly polymorphic, microsatellites have become one of the most popular genetic markers for making population genetic inferences in closely related populations.

Unequal crossing over (Richard and Pâques 2000; Smith 1976) and replication slippage (Levinson and Gutman 1987) are the two main mechanisms proposed to explain the high mutation rate of microsatellites. The simplest microsatellite model is the stepwise mutation model (SMM) proposed by Ohta and Kimura (1973), which states that the length of the microsatellite increases or decreases by one repeat unit at a rate independent of the microsatellite length. For SMM the Q matrix has the following entries:

$$q_{ij} = \begin{cases} \beta & \text{if } |i-j| = 1 \\ 0 & \text{if } |i-j| > 1 \\ -\sum_{k \neq i} q_{ik} & \text{if } i = j \end{cases}.$$

There are a large number of more complex models that have been introduced in the literature to account for length-dependent mutation rates, mutational bias (unequal rates of expansion and contraction) and 'two-phase' dynamics, in which the length of the repeat changes by more than one repeat unit with a single mutation. For a review of these models, and a description of a nested family of microsatellite models that encompasses most of the variants, see (Wu and Drummond 2011).

3.5   Felsenstein's likelihood

The phylogenetic likelihood of a time-tree g is the probability of the sequence data (D), given the phylogenetic tree and a substitution model. First consider a $n \times L$ matrix representing a sequence of length L at every node in the tree:

$$D = \begin{pmatrix} D_{I_1} \\ D_{I_2} \\ \vdots \\ D_{I_n} \end{pmatrix} = \begin{pmatrix} s_{I_1,1} & s_{I_1,2} & \cdots & s_{I_1,L} \\ s_{I_2,1} & s_{I_2,2} & \cdots & s_{I_2,L} \\ \vdots & \vdots & \ddots & \vdots \\ s_{I_n,1} & s_{I_n,2} & \cdots & s_{I_n,L} \end{pmatrix},$$

where I_k is the index of the kth leaf node and D_{I_k} is the sequence associated with it. The entry $s_{I_k,j}$ is a nucleotide base at site j of the sequence, taking values in the set $C = \{A, C, G, T\}$.

The likelihood of the tree g is:

$$L(g) = \Pr\{D|g, \Omega\},$$

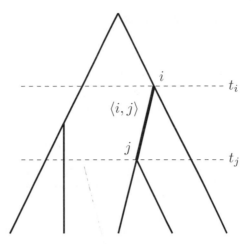

Figure 3.6 A focal branch $\langle i,j \rangle$ in the time-tree $g = \{R, \mathbf{t}\}$.

where $\Omega = \{Q, \mu\}$ includes parameters of the substitution model and the overall rate μ.

Let D_Y represent the $(n-1) \times L$ matrix whose rows are the unknown ancestral sequences at the internal nodes (Y) of the tree g:

$$D_Y = \begin{pmatrix} D_{Y_1} \\ D_{Y_2} \\ \vdots \\ D_{Y_{n-1}} \end{pmatrix} = \begin{pmatrix} s_{Y_1,1} & s_{Y_1,2} & \cdots & s_{Y_1,L} \\ s_{Y_2,1} & s_{Y_2,2} & \cdots & s_{Y_2,L} \\ \vdots & \vdots & \ddots & \vdots \\ s_{Y_{n-1},1} & s_{Y_{n-1},2} & \cdots & s_{Y_{n-1},L} \end{pmatrix}.$$

Let $D = C^{(n-1)L}$ denote the set of all possible ancestral sequences.

Consider an edge $\langle i,j \rangle \in R$ of tree $g = \{R, \mathbf{t}\}$ (see Figure 3.6). The individual associated with node j is a direct descendant of the individual associated with node i. However, the sequences D_i and D_j may differ if substitutions have occurred in the interval. Applying a CTMP model of substitution we have $\Pr\{s_{j,k} = c' | s_{i,k} = c\} = \left[e^{Q\mu(t_i - t_j)} \right]_{c,c'}$. Then the probability for any particular set of sequences D to be realised at the leaf nodes of the tree is:

$$\Pr\{D|g, \Omega\} = \sum_{D_Y \in D} \prod_{\langle i,j \rangle \in R} \prod_{k=1}^{L} \left[e^{Q\mu(t_i - t_j)} \right]_{s_{i,k}, s_{j,k}}.$$

(In the above formula, compact notation is obtained by including in the product over edges an edge terminating at the root from an ancestor of infinite age.)

The sum over all possible ancestral sequences D_Y looks onerous, but using a pruning algorithm Felsenstein (1981) demonstrated an efficient polynomial-time algorithm that makes this integration over unknown ancestral sequences feasible (see Section 3.7).

3.6   Rate variation across sites

It is common to allow rate variation across sites, and a key component of most models of rate variation across sites is the discrete gamma model introduced by Yang (1994). For K discrete categories this involves K times the computation as can be seen in the likelihood where an extra sum is used to average the likelihood over the K categories for each site in the alignment:

$$\Pr\{D|g, \Omega\} = \prod_{k=1}^{L} \sum_{c=1}^{K} \frac{1}{K} \left(\sum_{D_Y \in D} \prod_{\langle i,j \rangle \in R} \left[e^{Q\mu r_c(t_i - t_j)} \right]_{s_{i,k}, s_{j,k}} \right).$$

Here r_c is the relative rate of the cth rate category in the discrete gamma distribution and $\Omega = \{Q, \mu, \gamma\}$ includes the shape parameter (α) of the discrete gamma model of rate variation across sites, governing the values of $r_1, r_2, \ldots, r_K$. Figure 3.7 shows how the density of the continuous gamma distribution varies from L-shaped ($\alpha \leq 1$) to bell-shaped ($\alpha \gg 1$) with the shape parameter α.

Despite this flexibility, the introduction of a separate category of *invariant* sites that are assumed to have an evolutionary rate of zero can improve the fit to real data. This is the so-called $\Gamma + I$ approach to modelling rate variation (Gu et al. 1995; Waddell and Penny 1996). Model selection will often favour $\Gamma + I$ over Γ, although accurate estimation of the two parameters (a proportion of invariant sites, p_{inv} and α) is highly sensitive to sampling effects (Sullivan and Swofford 2001; Sullivan et al. 1999).

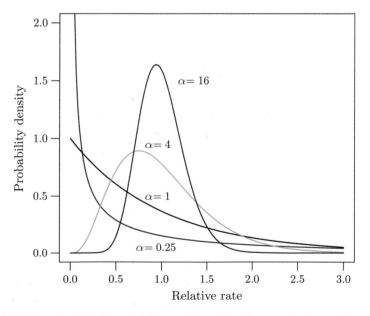

Figure 3.7 The gamma distribution for different values of the shape parameter α with scale parameter set to $1/\alpha$ so that the mean is 1 in all cases.

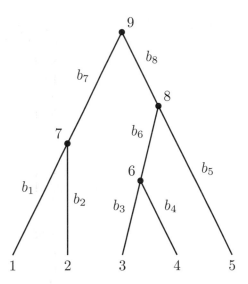

Figure 3.8 An example tree to illustrate the 'pruning' algorithm.

3.7   Felsenstein's pruning algorithm

The 'pruning' algorithm for computing the phylogenetic likelihood was introduced by Felsenstein (1981). In the following discussion we will consider a single site **s** and the corresponding nucleotide states associated with the ancestral nodes $\mathbf{s}_Y$ in the five-taxon tree in Figure 3.8:

$$\mathbf{s} = \begin{pmatrix} s_1 \\ s_2 \\ s_3 \\ s_4 \\ s_5 \end{pmatrix}, \quad \mathbf{s}_Y = \begin{pmatrix} s_6 \\ s_7 \\ s_8 \\ s_9 \end{pmatrix}.$$

Together they form a full observation:

$$\mathbf{s}_V = \begin{pmatrix} \mathbf{s} \\ \mathbf{s}_Y \end{pmatrix} = \begin{pmatrix} s_1 \\ s_2 \\ \vdots \\ s_9 \end{pmatrix}.$$

Consider the tree in Figure 3.8. If we had full knowledge of the sequences at internal nodes of this tree then the probability of site pattern $\mathbf{s}_V$ can be easily computed as a product of transition probabilities over all branches $\langle i, j \rangle$ multiplied by the prior probability of the nucleotide state at the root node (π_{s_9}), i.e.:

$$\Pr\{\mathbf{s}_V | g\} = \pi_{s_9} \prod_{\langle i,j \rangle} p_{s_i s_j}(b_j)$$

$$= \pi_{s_9} p_{s_9 s_7}(b_7) p_{s_7 s_1}(b_1) p_{s_7 s_2}(b_2)$$

$$\times p_{s_9 s_8}(b_8) p_{s_8 s_5}(b_5) p_{s_8 s_6}(b_6) p_{s_6 s_3}(b_3) p_{s_6 s_4}(b_4).$$

However, since we don't know the ancestral nucleotides s_Y we must integrate over their possible values to get the probability of the leaf data:

$$\Pr\{s|g\} = \sum_{s_Y} \Pr\{s_V|g\}$$

$$= \sum_{s_9 \in C} \sum_{s_8 \in C} \sum_{s_7 \in C} \sum_{s_6 \in C} \pi_{s_9} p_{s_9 s_7}(b_7) p_{s_7 s_1}(b_1) p_{s_7 s_2}(b_2)$$

$$\times p_{s_9 s_8}(b_8) p_{s_8 s_5}(b_5) p_{s_8 s_6}(b_6) p_{s_6 s_3}(b_3) p_{s_6 s_4}(b_4).$$

Felsenstein makes the point that you can move the summations in the above equation rightwards to reduce the amount of repeated calculation:

$$\Pr\{s|g\} = \sum_{s_9 \in C} \pi_{s_9} \left\{ \sum_{s_7 \in C} p_{s_9 s_7}(b_7) \left[p_{s_7 s_1}(b_1) \right] \left[p_{s_7 s_2}(b_2) \right] \right\}$$

$$\times \left\{ \sum_{s_8 \in C} p_{s_9 s_8}(b_8) \left[p_{s_8 s_5}(b_5) \right] \left[\sum_{s_6 \in C} p_{s_8 s_6}(b_6) \left(p_{s_6 s_3}(b_3) \right) \left(p_{s_6 s_4}(b_4) \right) \right] \right\}.$$

$$(3.2)$$

Notice that the pattern of brackets mirrors the topology of the tree. This is a clue that Equation (3.2) can be redefined in terms of a recursion that can be efficiently computed by dynamic programming.

Now define a matrix of partial likelihoods:

$$L = \begin{pmatrix} L_{1,A} & L_{1,C} & L_{1,G} & L_{1,T} \\ L_{2,A} & L_{2,C} & L_{2,G} & L_{2,T} \\ \vdots & \vdots & \vdots & \vdots \\ L_{2n-1,A} & L_{2n-1,C} & L_{2n-1,G} & L_{2n-1,T} \end{pmatrix},$$

where $L_{i,c}$ is the partial likelihood of the data under node i, given the ancestral character state at node i is $c \in C$. The entries of L can be defined recursively. Assuming the two descendant branches of internal node y are $\langle y, j \rangle$ and $\langle y, k \rangle$ we have:

$$L_{y,c} = \left[\sum_{x \in C} p_{cx}(b_j) L_{j,x} \right] \times \left[\sum_{x \in C} p_{cx}(b_k) L_{k,x} \right].$$

For leaf node i the partial likelihoods are simply:

$$L_{i,c} = \begin{cases} 1 & \text{if } c = s_i \\ 0 & \text{otherwise} \end{cases}.$$

From this formulation it is clear that the partial likelihoods in L can be computed by filling the table row by row proceeding from the top. Each entry takes time proportional to the number of character states $S = |C|$, so each row takes time $O(S^2)$. Thus the overall running time of the algorithm is $O(nS^2)$; linear in the number of taxa and quadratic in the number of possible character states.

3.8   Miscellanea

3.8.1   Model averaging

Instead of selecting a single substitution model for a fixed subset of sites, MCMC makes it possible to average over substitution models and sites. Using reversible jump, Huelsenbeck et al. (2004) average over all nucleotide substitution models. In contrast, Wu et al. (2013) and Bouckaert et al. (2013) use the same idea to average over just a hierarchy of substitution models (see Section 15.4 for the latter hierarchy). Wu et al. (2013) group sites and assign a substitution model and rate to each group, but no structure among sites is assumed, and the number and constitution of groups is averaged over during the MCMC run. On the other hand, Bouckaert et al. (2013) average over sites, but group sites in consecutive blocks. Both models are available in BEAST.

3.8.2   Models for language evolution

Languages can be encoded by binary sequences representing the presence or absence of cognates. Cognates are word forms that share a common ancestry, such as 'hand' in English, Dutch and German. In French, the word for hand is 'main' and in Spanish and Italian it is 'mano'. So the first three languages get a '1' for the 'hand' cognate and '0' for the 'main' form, while in the last three languages it is the other way around. Languages encoded this way give an alignment that lends itself to the same kind of analysis as DNA sequences (Bouckaert et al. 2012; Gray and Atkinson 2003; Gray et al. 2009). However, specialised substitution models are required to deal with these data. The most successful are the continuous-time Markov chain (CTMC) model (which is the equivalent of GTR for just two states), the covarion model (Penny et al. 2001) and the stochastic Dollo model (Nicholls and Gray 2006). The covarion model assumes a fast and slow rate of evolution and allows switching between the two. The stochastic Dollo model is inspired by Dollo's law, which states that cognates can evolve only once.

3.8.3   Substitution models for morphological data

Often, morphological data are available for the taxa in sequence alignments. The MK model (Lewis 2001) is a generalisation of the Jukes–Cantor model for different numbers of states; it uses a rate matrix with constant rates throughout, but the size of the matrix is adjusted to the number of values the morphological characters can take. Morphological data are typically selected based on the fact that there is variation in these features. The MKv model (Lewis 2001) takes this into account by extending the MK model by using ascertainment correction for constant sites. Using the MKv model instead of the MK model results in more accurate branch length estimates. This model was applied by Pyron (2011) to estimate divergence times of *Lissamphibia* by dividing morphological characters into groups with the same number of states.

4 The molecular clock

4.1 Time-trees and evolutionary rates

The dual concepts of a time-tree and a molecular clock are central to any attempt at interpreting the chronological context of molecular variation. As we saw in Chapter 2, several natural tree prior distributions (e.g. coalescent and birth–death families) generate time-trees, rather than unrooted trees. But how are these time-trees reconciled with the genetic differences between sequences modelled in Chapter 3? The answer is the application of a *molecular clock*. The concept of a molecular clock traces back at least to the 1960s. One of the early applications of the molecular clock was instrumental in a celebrated re-calibration of the evolutionary relatedness of humans to other great apes, when Allan Wilson and Vincent Sarich described an 'evolutionary clock' for albumin proteins and exploited the clock to date the common ancestor of humans and chimpanzees to five million years ago (Sarich and Wilson 1967).

4.2 The molecular clock

The molecular clock is not a metronome. Each tick of the clock occurs after a stochastic waiting time, determined by the substitution rate μ. Let $D(t)$ be a random variable representing the number of substitutions experienced over evolutionary time t. The probability that exactly k substitutions have been experienced in time t is:

$$Pr\{D(t) = k\} = \frac{e^{-\mu t}(\mu t)^k}{k!}.$$

Figure 4.1 shows six realisations of the Poisson accumulation process, $D(t)$, showing the total number of substitutions accumulated through time. All of the realisations have the same substitution rate ($\mu = 1$), but the time at which they reach 25 substitutions varies substantially. This figure was produced with the following R code:

```
# set substitution rate to 1.0
mu <- 1.0
# construct matrix with six poisson trajectories
# each column has 25 cumulating exponential waiting times
ti <- replicate(6, c(0,cumsum(rexp(25,mu))))
# construct a parallel matrix of unit steps
k <- replicate(ncol(ti), 1:nrow(ti)-1)
# plot each column of the matrix in unit steps
matplot(ti, k, type="s", xlab="t", ylab="D(t)", lty=1)
```

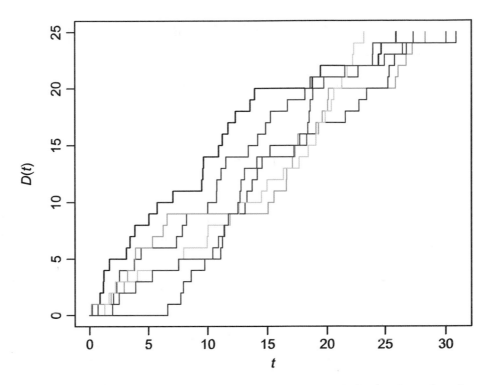

Figure 4.1 Six realisations of a Poisson accumulation process ($\mu = 1$) showing the total number of substitutions accumulated through time. All the realisations have the same substitution rate, but the time at which they reach 25 substitutions varies substantially.

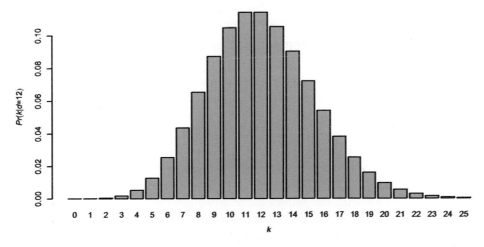

Figure 4.2 The Poisson probability distribution of experiencing k substitution events in genetic time $d = \mu t = 12$.

For example, if a substitution occurs, on average, once a week (say in a lineage of a rapidly evolving virus) then after 12 weeks the expected number of substitutions is 12. Figure 4.2 shows the wide range of actual outcomes possible around this expectation,

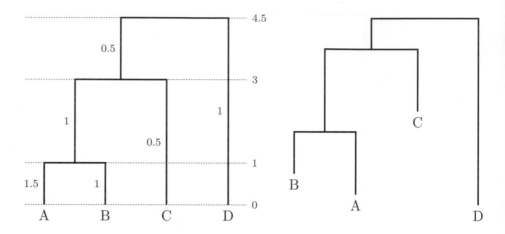

Figure 4.3 A time-tree of four taxa, branches labelled with rates of evolution and the resulting non-clock-like tree with branches drawn proportional to substitutions per site.

with 95% of such lineages having between 6 and 18 substitutions inclusive. The plot in this figure can be produced in R with the following two lines of code:

```
1  # Plot poisson distribution; k = 0 to 25 events, d=r*t=12
2  k <- 0:25; d <- 12;
3  barplot(dpois(k,d),names.arg=k,xlab="k",ylab="Pr(k|d=12)")
```

4.3   Relaxing the molecular clock

Researchers have grappled with the tension between molecular and non-molecular evidence for evolutionary time scales since Wilson and Sarich's ground-breaking work (Sarich and Wilson 1967; Wilson and Sarich 1969) on the ancestral relationship between humans and chimpanzees. Recently, a number of authors have developed 'relaxed molecular clock' methods. These methods accommodate variation in the rate of molecular evolution from lineage to lineage (see Figure 4.3 which depicts a time-tree of four taxa, branches labelled with varying rates of evolution, alongside a tree with branch lengths that display the resulting non-clock-like genetic distances). In addition to allowing non-clock-like relationships among sequences related by a phylogeny, modelling rate variation among lineages in a gene tree also enables researchers to incorporate multiple calibration points that may not be consistent with a strict molecular clock. These calibration points can be associated either with the internal nodes of the tree or the sampled sequences themselves. Furthermore, relaxed molecular clock models appear to fit real data better than either a strict molecular clock or the other extreme of no clock assumption at all. In spite of these successes, controversy still remains around the particular assumptions underlying some of the popular relaxed molecular

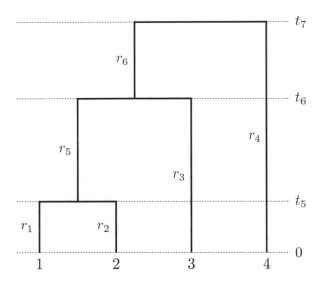

Figure 4.4 A time-tree of four taxa, branches labelled with rates of evolution.

clock models currently employed. A number of authors argue that changes in the rate of evolution do not necessarily occur smoothly nor on every branch of a gene tree. The alternative expounds that large subtrees share the same underlying rate of evolution and that any variation can be described entirely by the stochastic nature of the evolutionary process. These phylogenetic regions or subtrees of rate homogeneity are separated by changes in the rate of evolution. This alternative model may be especially important for gene trees that have dense taxon sampling, in which case there are potentially many short, closely related lineages, amongst which there is no reason *a priori* to assume differences in the underlying rate of substitution.

A Bayesian framework for allowing the substitution rate to vary across branches has the following structure:

$$f(g, \mathbf{r}, \theta | D) = \frac{\Pr\{D | g, \mathbf{r}, \theta\} f(\mathbf{r} | \theta, g) f(g | \theta) f(\theta)}{\Pr\{D\}}, \tag{4.1}$$

where $\mathbf{r}$ is a vector of substitutions rates, one for each branch in the tree (g) (Figure 4.4). For brevity, the vector θ represents all other parameters in the model, including the parameters of the relaxed-clock model that govern $\mathbf{r}$ (e.g. μ and S^2 in the case of log-normally distributed rates among branches), as well as the parameters of the substitution model and tree prior. As usual, D is the multiple sequence alignment. In the following sections we will survey a number of alternative approaches to relaxing the molecular clock. In most cases the term $f(\mathbf{r} | \theta, g)$ can be further broken down into a product of densities over all the branches in the tree:

$$f(\mathbf{r} | \theta, g) = \prod_{\langle i,j \rangle \in R} f(r_j | \Phi),$$

where j is a unique index associated with the tip-ward node of branch $\langle i, j \rangle$ in tree g.

4.3.1   Autocorrelated substitution rates across branches

A model of autocorrelated substitution rates across branches was first introduced by Thorne et al. (1998). They employed a geometric Brownian motion model to specify $f(\mathbf{r}|\theta, g)$ in Equation (4.1), as follows:

$$f(r_i|\mu_R, \sigma^2) = \frac{1}{r_i\sqrt{2\pi\,\tau_i\sigma^2}} \exp\left\{-\frac{1}{2\tau_i\sigma^2}\left(\ln(r_i/r_{A(i)})\right)^2\right\},$$

where $A(i)$ is the index of the parent branch of the ith branch, $r_{A(root)} = \mu_R$ by definition, and τ_i is the time separating branch i from its parent. The precise definition of τ_i and the means by which the r_i parameters are applied to the tree has varied in the literature. In the original paper τ_i was defined as the time between the midpoint of branch i and the midpoint of its parent branch:

$$\tau_i = t_{A(i)} + b_{A(i)}/2 - (t_i + b_i/2),$$

where t_i is the time of the tip-ward node of the ith branch and $b_i = t_{A(i)} - t_i$.

Subsequently, Kishino et al. (2001) associated a rate r_i' with the ith *node* (the node on the tip-ward end of the branch i, whose time is t_i) and applied a *bias-corrected* geometric Brownian motion model:

$$f(r_i'|\mu_R, \sigma^2) = \frac{1}{r_i'\sqrt{2\pi\,b_i\sigma^2}} \exp\left\{-\frac{1}{2b_i\sigma^2}\left(\ln(r_i'/r_{A(i)}') + \frac{b_i\sigma^2}{2}\right)^2\right\}.$$

The aim of the bias correction term $\frac{b_i\sigma^2}{2}$ is to produce a child substitution rate whose expectation is equal to that of the parent rate, i.e. $E(r_i) = r_{A(i)}$. Finally, instead of using r_i' as the substitution rate of the ith branch, Kishino et al. (2001) used the arithmetic average of the two node-associated rates: $r_i = (r_i' + r_{A(i)}')/2$.

There are many variations on the autocorrelated clock models described above, including the compound Poisson process (Huelsenbeck ct al. 2000), and the Thorne–Kishino continuous autocorrelated model (Thorne and Kishino 2002). The first draws the rate multiplier r_i from a Poission distribution, while the latter draws r_i from a log-normal with mean $r_{A(i)}$ and variance proportional to the branch length.

4.3.2   Uncorrelated substitution rates across branches

The basic idea behind the 'uncorrelated rates across branches' model is that *a priori* substitution rates are independent and identically distributed across branches. The ith branch will have a rate r_i drawn independently from the *parent* distribution. Drummond et al. (2006) considered the exponential and log-normal distributions for parent distributions. Rannala and Yang (2007) also considered the log-normal distribution. The log-normal distribution has the benefit of including a parameter governing the variance of rates across branches and thus allows a measure of the extent to which the strict clock requires relaxing. The probability density of r_i, the rate of the

ith branch, is given by:

$$f(r_i|M, S) = \frac{1}{r_i\sqrt{2\pi S^2}} \exp\left\{-\frac{1}{2S^2}[\ln(r_i) - M]^2\right\},$$

where M is the mean of the logarithm of the rate and S^2 is the variance of the log rate. An alternative parameterisation used by Rannala and Yang (both parameterisations are available in BEAST) employs the mean rate μ instead of M:

$$f(r_i|\mu, S) = \frac{1}{r_i\sqrt{2\pi S^2}} \exp\left\{-\frac{1}{2S^2}[\ln(r_i/\mu) + 1/2S^2]^2\right\}.$$

Regardless of the parameterisation, a number of technical alternatives to sampling such a model in an MCMC framework have been described (Drummond et al. 2006; Li and Drummond 2012; Rannala and Yang 2007). The independent gamma rate model (Lepage et al. 2007) is similar to the uncorrelated clock model but draws clock rates from a gamma distribution with variances proportional to the branch length.

4.3.3 Random local molecular clocks

Local molecular clocks are another alternative to the strict molecular clock (Yoder and Yang 2000). A local molecular clock permits different regions in the tree to have different rates, but within each region the rate must be the same. Early applications of local molecular clock models did not include the modelling of uncertainty in the (1) phylogenetic tree topology, (2) the number of local clocks or (3) the phylogenetic positions of the rate changes between the local clocks. However, more recently a Bayesian approach has been described that allows averaging over all possible local clock models in a single analysis (Drummond and Suchard 2010). For a model that allows one rate change on a rooted tree there are $2n - 2$ branches on which the rate change can occur. To consider two rate changes, one must consider $(2n - 2) \times (2n - 3)$ possible rate placements. If each branch can have 0 or 1 rate changes then the total number of local clock models that might be considered is 2^{2n-2}, where n is the number of taxa in the tree. For even moderate n this number of local clock models cannot be evaluated exhaustively.

The Bayesian random local clock (RLC) model (Drummond and Suchard 2010) nests all possible local clock configurations, and thus samples a state space that includes the product of all 2^{2n-2} possible local clock models on all possible rooted trees. Because the RLC model includes the possibility of zero rate changes, it also serves to test whether one rate is sufficient, or whether some relaxation of a single molecular clock is required for the phylogeny at hand.

The RLC model comes in two flavours. Both are parameterised by a substitution rate associated with the branch above the root (r_{root}) and a pair of vectors, $\Phi = \{\phi_1, \phi_2, \ldots, \phi_{2n-2}\}$ and $\mathbf{B} = \{\beta_1, \beta_2, \ldots, \beta_{2n-2}\}$.

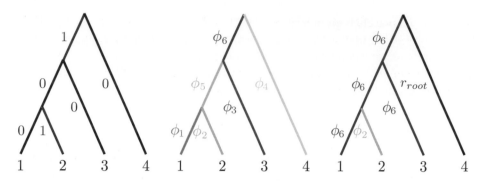

Figure 4.5 Left, the indicator variables for changes in rate. Centre, branches labelled with corresponding elements of Φ. Right, the resulting rates on the branches, once the indicator mask is applied.

The simpler of the two flavours of random local clock computes branch rates for all nodes apart from the root as follows:

$$r_i = \begin{cases} r_{A(i)} & \text{if } \beta_i = 0 \\ \phi_i & \text{otherwise} \end{cases}.$$

From this it can be seen that **B** acts as a set of binary indicators, one for each branch. If $\beta_i = 0$ then the ith branch simply inherits the same substitution rate as its parent branch, otherwise it takes up an entirely independent rate, ϕ_i. Thus the branches where $\beta_i = 1$ are the locations in the tree where the substitution rate changes from one local clock to the next (see Figure 4.5). Setting all β_i to zero gives rise to a strict clock.

The second flavour, and the one described in detail by Drummond and Suchard (2010) computes branch lengths as follows:

$$r_i = \begin{cases} r_{A(i)} & \text{if } \beta_i = 0 \\ \phi_i \times r_{A(i)} & \text{otherwise} \end{cases}.$$

This model retains some autocorrelation in substitution rate from a parent local clock to a 'child' local clock, with the parameters of Φ interpreted as relative rate modifiers in this framework, rather than absolute rates as in the first construction. Regardless, in order to sample either of these models in a Bayesian framework, the key to success is the construction of an appropriate prior on Φ and **B**. Drummond and Suchard (2010) defined K, the number of rate change indicators:

$$K = \sum_{i=1}^{2n-2} \beta_i$$

and suggested a prior on **B** that would induce a truncated Poisson prior distribution on K:

$$K \sim \text{Truncated-Poisson}(\lambda).$$

The prior on Φ can be chosen relatively freely. Independent and identically distributed from a log-normal distribution would be quite appropriate for the elements of Φ in the uncorrelated RLC model. Independent and identically distributed from a gamma distribution has been suggested for elements of Φ in the autocorrelated RLC model (Drummond and Suchard 2010).

Although the basic idea of the random local clock model is simple, there are a number of complications to its implementation in a Bayesian framework that are not dealt with here. These include normalisation of the substitution rate across the phylogeny in the absence of calibrations and issues related to sampling across models of differing dimensions. For readers interested in these finer details, we refer you to the original paper by Drummond and Suchard (2010).

4.3.4 Branch rates vs. site rates

In Section 3.6 we looked at calculating the likelihood of the data $\Pr\{D|g, \Omega\}$ with variation of rates among sites, typically implemented by averaging over various categories with a different rate for each category. When calculating $\Pr\{D|g, \Omega\}$ both branch and site rates are involved. For a particular category, a site rate is a multiplier for the complete tree, that is, every branch length is multiplied by the site rate. Furthermore, every branch length is multiplied by the branch rate, and this branch rate is constant over the various categories. So, for a certain rate category r_c the length of branch i is multiplied by both r_c and branch rate r_i, but the rate category r_c varies with different categories, while the branch rate r_i can vary with different branches.

4.4 Calibrating the molecular clock

Molecular clock analysis and divergence-time dating go hand in hand. The calibration of a molecular clock to an absolute time scale leads directly to ages of divergences in the associated time-tree. Although divergence-time dating is a well-established cornerstone of evolutionary biology, there is still no widely accepted objective methodology for converting information from the fossil record to calibration information for use in molecular phylogenies.

4.4.1 Node dating

A standard, though well-criticised (for a critique see Ronquist et al. 2012b) approach to divergence-time dating has been termed *node dating* by Ronquist et al. (2012b). It involves applying the geological age of a fossil as a 'calibration' to the phylogenetic divergence it is deemed to correspond to. Usually the geological age of the fossil is transformed into a *calibration density* aimed at capturing the various sources of uncertainty inherent in associating a fossil with a particular divergence.

There are a number of problems with this approach. The three most serious shortcomings of node dating are (1) the reliance on indirect use of the fossil evidence to determine which molecular divergence the fossil corresponds to, (2) the need to determine the exact form of the calibration density and (3) the assumption that the fossil does indeed correspond to a node in the molecular phylogeny (as opposed to some extinct offshoot, or to a direct ancestor along one of the lineages in between two divergences). These shortcomings are somewhat intertwined and the last two can be somewhat alleviated by developing a calibration density for the dated node based on an explicit model of phylogenetic speciation and extinction. Using stochastic simulation based on a simple phylogenetic birth–death model, Monte Carlo model-based calibration densities can be constructed (Matschiner and Bouckaert 2013).

However, regardless of how the calibration density for a phylogenetic divergence is arrived at, there is an additional challenge for Bayesian node dating. The task of correctly constructing a Bayesian tree prior that composes one or more fossil calibration densities with an underlying tree process prior involves computational challenges when the phylogenetic tree is also inferred (Heled and Drummond 2012, 2013).[1]

4.4.2   Leaf dating

For some types of molecular data, the difference in sampling dates of the sequenced taxa is sufficient to provide direct calibration of the molecular clock (Drummond et al. 2002; Rambaut 2000). Such molecular data is variously termed serially sampled, time-stamped or heterochronous sequence data and can be said to come from *measurably evolving populations* (Drummond et al. 2003a). The principal sources of such data are ancient DNA (e.g. Lambert et al. 2002; Shapiro et al. 2004) and rapidly evolving pathogens (Kühnert et al. 2011), including RNA viruses (for a review see Jenkins et al. 2002) and whole-genome sequenced bacteria (e.g. Ford et al. 2011).

4.4.3   Total-evidence leaf dating

A recent attempt to provide a principled alternative to node dating was described by the authors as 'a total-evidence approach to dating with fossils' (Ronquist et al. 2012b). This approach essentially uses *leaf dating* of fossil taxa, but relies on morphological data instead of molecular data to position the fossil taxa in the phylogeny. Each fossil is represented by a leaf in a (total-evidence) time-tree, that also includes leaves representing all extant species. The age of each fossil leaf is determined by the geological stratum in which the fossil was discovered. The uncertainty in the phylogenetic location of the fossil leaves are integrated over via MCMC using the phylogenetic likelihood of morphological data collected from both extant and fossil taxa. If available, molecular data can also contribute to the estimation of the phylogeny connecting the extant taxa. This model thus addresses the first shortcoming of node dating by directly *estimating* the phylogenetic placement of fossils from coded morphological data. It also somewhat obviates the second shortcoming, since a large component of the uncertainty

---

[1] The composition of calibrated birth–death tree priors is more straightforward for Bayesian divergence-time dating on a fixed tree topology (Yang and Rannala 2006).

in a calibration density aims to describe the unknown time separating the fossil from the divergence it is associated with. That separation time is effectively directly modelled in total-evidence dating. However, Ronquist et al.'s approach does not fully address the final shortcoming of node dating because their model effectively posits that all fossils are 'extinct offshoots', by representing them as leaves in the phylogeny.[2]

4.4.4 Total-evidence dating with fossilised birth–death tree prior

The recently introduced fossilised birth–death (FBD) process (Heath et al. 2014) aims at a promising refinement of total-evidence dating. But using the FBD model as a tree prior effectively requires a new class of phylogeny – *sampled ancestor trees* – to allow for fossil taxa to fall into two categories:

1. Fossil taxa that are direct ancestors of other taxa (fossil or extant) in the total-evidence phylogeny.
2. Fossil taxa that represent 'extinct offshoots', with no direct descendants among the other taxa (fossil or extant).

To exploit the FBD model as a tree prior for Bayesian total-evidence dating, a new Bayesian MCMC algorithm that can sample the space of sampled ancestor trees is required and has been recently developed (Gavryushkina et al. 2014). Using this Bayesian sampling algorithm (Gavryushkina et al. 2014) along with a substitution model for morphological data would allow for simultaneously estimation of a total-evidence phylogenetic tree and its divergence times while also providing automatic classification of all fossil taxa into the two categories above. Such a framework would address all the current shortcomings of Bayesian node dating.

---

[2] Although one could perhaps consider a fossil to be a direct ancestor of another leaf if the posterior estimate of the branch length attaching it to the phylogeny was very short.

5 Structured trees and phylogeography

This chapter describes multi-type trees and various extensions to the basic phylogenetic model that can account for population structure, geographical heterogeneity and epidemiological population dynamics.

5.1 Statistical phylogeography

Phylogeography can be viewed as an approach that brings together phylogenetics and biogeography (Avise 2000). Phylogeography has a long historyand the methods employed are very diverse. Phylogeographical patterns can be explored within a single species or between closely related species, and these patterns can be used to address questions of geographical origins and expansions, island biogeography (e.g. Canary Islands (Sanmartín et al. 2008)), range expansions/contractions and effects of environmental and climate changes on geographical dispersal and extent. As with other chapters in this book, we don't attempt to be comprehensive, but instead point to some relevant material and focus on approaches that we are familiar with and believe to have promise. We mainly consider methods that attempt to directly reconcile geographic data with the phylogenetic relationships of the sampled taxa.[1] Until recently, due to its simplicity, the most popular method reconciling discrete geographical locations with phylogenetic relationships (and as a result inferring ancestral locations on the phylogenetic tree) was maximum parsimony (Maddison and Maddison 2005; Slatkin and Maddison 1989; Swofford 2003; Wallace et al. 2007). However, this method doesn't allow for a probabilistic assessment of the uncertainty associated with the reconstruction of ancestral locations. As with the field of phylogenetics, phylogeography recently experienced a transition to model-based inference approaches. When it finally came, the struggle for authority between model-based approaches and the alternatives[2] was relatively short compared to the protracted 'troubled growth of statistical phylogenetics' recounted by Felsenstein (2001). History will probably find that the argument in defence of model-based approaches for phylogeography mounted in the journal

---

[1] But note that in some methods described below the sampled taxa may be explicitly recognised as a statistical sample from underlying populations, of which properties are inferred, such as levels of gene flow, or geographic isolation.

[2] Alternatives to model-based statistical phylogeography are mainly forms of parsimony in various guises, sometimes ironically called 'statistical phylogeography' or 'statistical parsimony'.

Molecular Ecology by Mark Beaumont and a swath of fellow proponents of Approximate Bayesian Computation (ABC) and other forms of model-based inference (Beaumont et al. 2010) was the decisive moment. That paper probably marks the analogous transition in phylogeographic research to the transition made in phylogenetic research a decade or more before. We will only consider model-based methods in this chapter. Statistical methods for phylogeography include those that perform inferences conditional on a tree, and those that jointly estimate the phylogeny, the phylogeographic state and the parameters of interest. Most of the models described below could be applied in either context. Although we mainly discuss models for which full Bayesian inference is already feasible, it is worth noting that ABC has had a large role in the development of model-based phylogeographic inference (Beaumont et al. 2010). When full Bayesian approaches aren't available, then well-implemented ABC approaches using a good choice of summary statistics can be effective. However, when a full Bayesian implementation is available for the model of interest we would always prefer it, since it will be a lower variance estimator. In the next sections we will focus on description of the various models and the necessary augmentation of the time-tree to allow inference under them. Most of these models are conceptually united by employing inference on multi-type branching processes.

5.2 Multi-type trees

We define a multi-type tree $\mathcal{T}$ of n leaves as a fully resolved time-tree in which every internal node represents a coalescence/divergence and where every point on each edge of the tree is associated with exactly one type s drawn from a fixed set S of such types. In the context of phylogeography or geographically structured population genetics, these types can be considered as a set of discrete geographical locations or demes. However, the mathematical description is quite general. Models that describe (marginal) probability distributions on multi-type trees arise when considering a diverse range of evolutionary phenomena, including (neutral) molecular evolution (Felsenstein 1981), migration (Beerli and Felsenstein 2001; Ewing et al. 2004; Lemey et al. 2009a), single locus non-neutral population genetics (Coop and Griffiths 2004), trait-driven speciation/extinction (Maddison 2007) and compartmental epidemiological dynamics (Palmer et al. 2013; Volz 2012).

Mathematically, we write $\mathcal{T} = (V, R, \mathbf{t}, M)$. The first three elements are the usual phylogenetic tree components (defined in Table 2.1): a set of $2n - 1$ nodes V, a set R containing directed edges of the form $\langle i, j \rangle$ between nodes $i, j \in V$ and a set of node ages $\mathbf{t} = \{t_i | i \in V\}$ where t_i is the age of node i. The set of nodes is partitioned into two smaller sets Y and I representing the $n - 1$ internal and n external nodes, respectively. Every node $i \in V$ besides the root r has exactly one parent node i_p satisfying $t_{i_p} > t_i$. Each internal node $i \in Y$ has two child nodes i_{cl} and i_{cr} satisfying $t_{i_{cl}} < t_i$ and $t_{i_{cr}} < t_i$. The final element in $\mathcal{T}$ is unique to multi-type trees and is defined by $M = \{\varphi_{\langle i,j \rangle} | \langle i, j \rangle \in E\}$, where each function $\varphi_{\langle i,j \rangle} : [t_i, t_j] \to D$ is piecewise constant

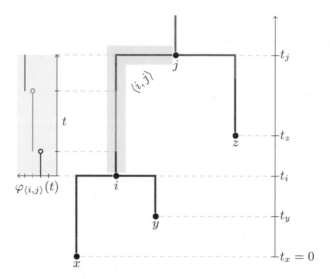

Figure 5.1 A multi-type tree $\mathcal{T} = (V, R, \mathbf{t}, M)$ with $V = I \cup Y$ where $I = \{x, y, z\}$, $Y = \{i, j\}$, $R = \{\langle x, i \rangle, \langle y, i \rangle, \langle i, j \rangle, \langle z, j \rangle\}$ and the coalescence times $\mathbf{t}$ and type mappings M are as shown. Here we have selected the type set $D = \{\mathbf{blue}, \mathbf{red}, \mathbf{green}, \mathbf{orange}\}$, although this can be composed of the values of any discrete trait.

and defined such that $\varphi_{\langle i,j \rangle}(t)$ is the type associated with the time t on edge $\langle i, j \rangle \in R$. Such a tree is illustrated in Figure 5.1.

5.2.1 Bayesian inference of multi-type trees

There are a variety of approaches to inference of multi-type trees and their associated parameters. The approach taken is generally dependent on two factors: (1) the statistical structure of the multi-type tree model and (2) the aspects of the multi-type tree model to be inferred. Some multi-type tree models assume independence of the multi-type process on individual edges of the tree (e.g. the *migration* model described in the following section). This sort of independence structure can be exploited to enable efficient integration over the number and times of the type-change events, conditional on the types at the tips of a fixed time-tree, using the phylogenetic likelihood (Felsenstein 1981). This is especially useful when the full history of type-changes is a nuisance, and the parameter of interest is, for example, the associated migration matrix. Whereas models that don't assume independence of the branching and type-changing processes must generally employ MCMC inference on the full multi-type tree state space,[3] including a random number of type transition events on the time-tree. This can be computationally intensive, but there have been some recent efforts to improve the computational efficiency of such

---

[3] In analyses in which only the transition rate matrix needs to be inferred (without the full multi-type tree) it is possible to numerically integrate over all possible transition events, which may be computationally less intensive (e.g. Stadler and Bonhoeffer 2013).

algorithms (Vaughan et al. 2014). More details are provided in the description of the individual models below.

5.3   Mugration models

A *mugration model* is a mutation or substitution model used to analyse a migration process. A recent study of influenza A H5N1 virus introduced a fully probabilistic 'mugration' approach by modelling the process of geographic movement of viral lineages via a continuous-time Markov process where the state space consists of the locations from which the sequences have been sampled (Lemey et al. 2009a). This facilitates the estimation of migration rates between pairs of locations. Furthermore, the method estimates ancestral locations for internal nodes in the tree and employs Bayesian stochastic variable selection (BSVS) to infer the dominant migration routes and provide model averaging over uncertainty in the connectivity between different locations (or host populations). This method has helped with the investigation of the influenza A H5N1 origin and the paths of its global spread (Lemey et al. 2009a), and also the reconstruction of the initial spread of the novel human influenza A H1N1 pandemic (Lemey et al. 2009b). However, the mugration models are limited to reconstructing ancestral locations from among a pre-defined set of locations usually made up of the sampled locations. As demonstrated by the analysis of the data set on rabies in dogs in West and Central Africa, absence of sequences sampled close to the root can hinder the accurate estimation of viral geographic origins (Lemey et al. 2009a). Phylogeographic estimation is therefore improved by increasing both the spatial density and the temporal depth of sampling. However, dense geographic sampling leads to large phylogenies and computationally intensive analyses. Increasingly large phylogenetic analyses have led to the development of phylogenetic likelihood implementations that can take advantage of the large number of processing cores on modern graphics processing units (GPUs) (Suchard and Rambaut 2009).

5.4   The structured coalescent

The *structured coalescent* (Hudson 1990) can also be employed to study phylogeography. The structured coalescent has also been extended to heterochronous data (Ewing et al. 2004), thus allowing the estimation of migration rates between demes in calendar units (see Figure 5.2). The *serial structured coalescent* was first applied to an HIV data set with two demes to study the dynamics of subpopulations within a patient (Ewing et al. 2004), but the same type of inference can be made at the level of the host population. Further development of the model allowed for the number of demes to change over time (Ewing and Rodrigo 2006a). MIGRATE (Beerli and Felsenstein 2001) also employs the structured coalescent to estimate subpopulation sizes and migration rates in both Bayesian and maximum likelihood frameworks and has also recently been used to investigate spatial characteristics of viral epidemics (Bedford et al. 2010).

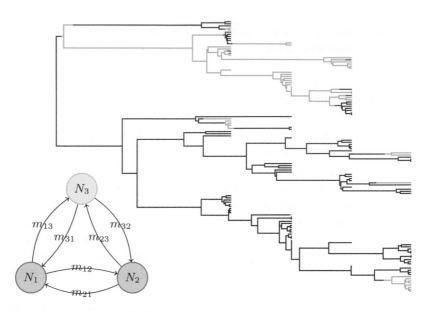

Figure 5.2 A simulation of the serially sampled structured coalescent on three demes. The population size of the three demes is equal ($N_1 = N_2 = N_3 = 1000$).

Additionally, some studies have focused on the effect of ghost demes (Beerli 2004; Ewing and Rodrigo 2006b). Although the structured coalescent model is promising, its application in Bayesian MCMC is computationally demanding because the standard form of the likelihood calculation (Beerli and Felsenstein 2001; Hudson 1990) requires that the genealogical tree be augmented with all of the unknown migration events in the ancestry of the sample. The migration events themselves are typically treated as nuisance parameters and integrated out using MCMC (Beerli and Felsenstein 2001; Ewing et al. 2004). Recently some effort has been made to apply *uniformisation* (Fearnhead and Sherlock 2006; Rodrigue et al. 2008) to obtain efficient Bayesian MCMC sampling algorithms for structured coalescent inference from large serially sampled data sets (Vaughan et al. 2014). However, despite this activity, to our knowledge there are no models explicitly incorporating population structure, heterochronous samples and non-parametric population size history yet available.

One *ad hoc* solution is the *mugration* model described in the previous section, which involves modelling the migration process along the tree in a way that is conditionally independent of the population sizes estimated by the skyline plot (Lemey et al. 2009a). Thus, conditional on the tree, the migration process is independent of the coalescent prior. However, this approach does not capture the interaction between migration and coalescence that is implicit in the structured coalescent, since coalescence rates should depend on the population size of the deme the lineages are in, leading to a natural interaction between the migration and branching processes. The *mugration* method also does not permit the population sizes of the individual demes to be accurately estimated as part of the analysis. As we will see in the following section, statistical phylogeography is one area where the unification of phylogenetic and mathematical epidemiological models looks very promising.

5.5   Structured birth–death models

The birth–death models introduced in Section 2.4 can also be extended to model population structure (Kendall 1949). Similarly to the structured coalescent process this results in a fully probabilistic approach in which the migration process among discrete demes depends on the characteristics of the demes.

Such multi-type birth–death models come in different flavours, depending on the research question posed. When samples are indeed taken from geographical locations with migration among them, migration events should be occurring along the branches in the phylogeny. In other cases type changes at branching events are more reasonable, e.g. when trying to identify superspreaders in an HIV epidemic (Stadler and Bonhoeffer 2013).

In either case, one can either employ the multi-type trees introduced above (see Figure 5.3), or integrate out the migration events such that standard BEAST trees can be used for inference of the migration rates among types.

When applied to virus epidemics, a birth–death tree prior allows the reconstruction of epidemiological parameters such as the effective reproduction number R (see Section 5.7). Using a structured birth–death model, these parameters can differ among demes and be estimated separately.

5.6   Phylogeography in a spatial continuum

In some cases it's more appropriate to model the spatial aspect of the samples as a continuous variable. The phylogeography of wildlife host populations have often been modelled in a spatial continuum by using diffusion models, since viral spread and host movement tend to be poorly modelled by a small number of discrete demes. One example is the expansion of geographic range in the eastern United States of the raccoon-specific rabies virus (Biek et al. 2007; Lemey et al. 2010). Brownian diffusion, via the comparative method (Felsenstein 1985; Harvey and Pagel 1991), has been utilised to model the phylogeography of feline immunodeficiency virus collected from the cougar (*Puma concolor*) population around western Montana. The resulting phylogeographic reconstruction was used as a proxy for the host demographic history and population structure due to the predominantly vertical transmission of the virus (Biek et al. 2006). However, one of the assumptions of Brownian diffusion is rate homogeneity on all branches. This assumption can be relaxed by extending the concept of relaxed clock models (Drummond et al. 2006) to the diffusion process (Lemey et al. 2010). Simulations show that the relaxed diffusion model has better coverage and statistical efficiency over Brownian diffusion when the underlying process of spatial movement resembles an over-dispersed random walk.

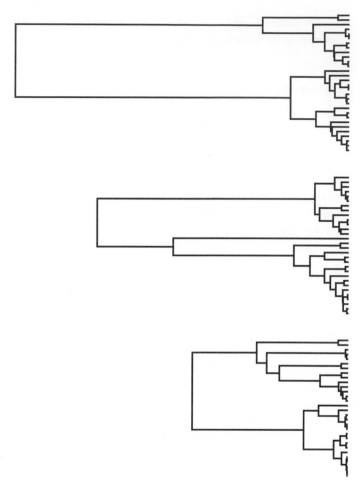

Figure 5.3 Three realisations of the structured coalescent on two demes. The population size of the two demes is equal ($N_0 = N_1 = 1000$) and the migration rates in both directions are $m_{01} = m_{10} = 0.00025$ in units of expected migrants per generation.

Like their *mugration model* counterparts, these models ignore the interaction of population density and geographic spread in shaping the sample genealogy. However, there has been progress in the development of mathematical theory that extends the coalescent framework to a spatial continuum (Barton et al. 2002, 2010a, 2010b), although no methods have yet been developed providing inference under these models.

5.7   Phylodynamics with structured trees

A new area, known as *phylodynamics* (Grenfell et al. 2004; Holmes and Grenfell 2009), promises to synthesise the distinct fields of mathematical epidemiology and statistical phylogenetics (Drummond and Rambaut 2007; Drummond et al. 2002, 2012; Ronquist et al. 2012a; Stadler 2010) to produce a coherent framework (Kühnert et al. 2014; Leventhal et al. 2014; Mollentze et al. 2014; Palmer et al. 2013; Rasmussen et al. 2011; Stadler and Bonhoeffer 2013; Stadler et al. 2013; Volz 2012; Volz et al. 2009; Welch 2011) in which genomic data from epidemic pathogens can directly inform sophisticated epidemiological models. Phylodynamics is particularly well suited to inferential epidemiology because many viral and bacterial pathogens (Gray et al. 2011) evolve so quickly that their evolution can be directly observed over weeks, months or years (Kühnert et al. 2011; Pybus and Rambaut 2009; Volz et al. 2013). So far, only part of the promise of phylodynamics has been realised. Early efforts include: (1) modelling the size of the pathogen population through time using a deterministic model for the epidemic (Volz 2012; Volz et al. 2009); (2) adopting new types of model for the transmission tree itself that are more suited to the ways in which pathogens are spread and sampled (Stadler et al. 2013); and (3) coupling this with an approximation to a stochastic compartmental model for the pathogen population (Kühnert et al. 2014; Leventhal et al. 2014). Only the last two of these approaches have been implemented in software and made available to practitioners. These efforts are just scratching the surface of this complex problem. They all make approximations that introduce biases of currently unknown magnitude into estimates.

Figure 5.4 depicts a multi-type (or structured) SIR process, in which there are two (coupled) demes, each of which is undergoing a stochastic SIR process. The number of infected individuals in the two demes is shown as a stochastic jump process, and the vertical lines emphasise the correspondence between events in the tree and events in the underlying infected populations. Every internal node in the tree corresponds to an infection event in the local epidemic of one of the demes. Likewise every migration

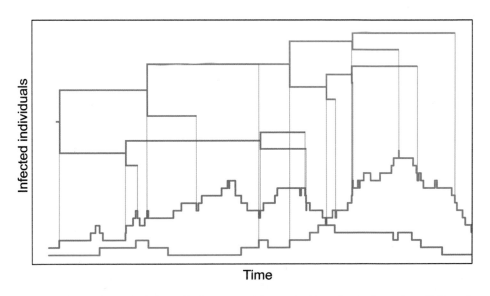

Figure 5.4 A two-deme phylodynamic time-tree with associated stochastic dynamics of infected compartments. (With thanks to Tim Vaughan for producing this figure.)

between the two demes corresponds with a simultaneous increment/decrement of the recipient/donor infected populations. It is easy to write down the likelihood of this model when both the multi-type tree and the stochastic trajectories are available. This implies that Bayesian inference is also available if the MCMC state space is augmented with the epidemic trajectories of infected individuals. However, these trajectories are large objects of random size (their size is determined by the number of infections, recoveries and migrations they represent). Efficient inference under these models may be limited to small outbreaks unless more efficient means to calculate the likelihood of multi-type trees under this model are discovered.

5.8   Conclusion

In this chapter we have outlined various approaches to modelling population structure and population dynamics in the context of phylogenetic time-trees. Most forms of structure entail each branch being associated, at each point in time, with one specific state from a discrete set of *types*. Transitions between types along a branch in the multi-type time-tree form a jump process. In most cases, branches of the same type, at the same time, are dynamically equivalent. In some models, the rates of the jump process along a branch in the tree are independent of the types of all other branches in the tree (e.g. Lemey et al. 2009a). In other models, the rates of the jump process are dependent on the states of other branches that exist in the same time in the tree (e.g. structured coalescent). Finally, in the most complex models described here, the rates of the jump process co-evolve with coupled stochastic processes (like the epidemic trajectory of the corresponding infected population in the structured SIR branching process, or the allele frequencies at the locus under selection in non-neutral coalescent models (Coop and Griffiths 2004)). We have not attempted a comprehensive review of these models because (apart from some notable exceptions) the application of Bayesian inference to this broad class of models is still in its infancy and it is an active and fast-moving area of research.

Part II

Practice

6 Bayesian evolutionary analysis by sampling trees

Molecular sequences, morphological measurements, geographic distributions, fossil and subfossil remains all provide a wealth of potential information about the evolutionary history of life on Earth, and the processes that have generated Earth's biodiversity. One of the challenges of modern evolutionary biology is the integration of these different data sources to address evolutionary hypotheses over the full range of spatial and temporal scales. Evolutionary biology has seen a transition from being a descriptive, mathematical science to a statistical and computational science. This transformation began first from an explosion of molecular sequence data with a parallel development of computational tools for their analysis. However, increasingly this transformation can be observed in other aspects of evolutionary biology where large global databases of other sources of information, such as fossils, geographical distributions and population history, are being curated and made publicly available.

The major goal of the developers of BEAST has been to design and build software and programming libraries that enable integration and statistical analysis of all these heterogeneous data sources. After years of development BEAST is now a robust and popular open source platform (http://beast2.org). The BEAST software is a popular Markov chain Monte Carlo (MCMC) sampler for phylogenetic models and has been downloaded tens of thousands of times. The program website for version 2 is http://beast2.org. The BEAST users' group (http://groups.google.com/group/beast-users) had over 2300 members as of July 2014. The original BEAST software article (Drummond and Rambaut 2007) is the most viewed paper in *BMC Evolutionary Biology* of all time (>68 000 views).

An underappreciated role in modern scientific research is the production of high-quality and robust software systems that provide for data analysis, hypothesis testing and parameter estimation. Typically development of such software was traditionally done without major support from research funding, because it is perceived to be a technical or engineering task that does not directly increase our scientific knowledge. However, we think this view is short-sighted as software like BEAST is an enabling technology that improves the level of statistical sophistication in evolutionary analyses and enables researchers to pose questions that could not previously be formally asked of the available data.

This chapter introduces the BEAST software for Bayesian evolutionary analysis through a simple step-by-step exercise. The exercise involves co-estimation of a gene phylogeny of primates and associated divergence times in the presence of calibration information from fossil evidence and illustrates some practical issues for setting up an analysis.

To run through this exercise, you will need the following software at your disposal, which is useful for most BEAST analyses:

- **BEAST** – this package contains the BEAST program, BEAUti, TreeAnnotator, DensiTree and other utility programs. This exercise is written for BEAST v2.2, which has support for multiple partitions. It is available for download from http://beast2.org/.
- **Tracer** – this program is used to explore the output of BEAST (and other Bayesian MCMC programs). It graphically and quantitatively summarises the distributions of continuous parameters and provides diagnostic information. At the time of writing, the current version is v1.6. It is available for download from http://tree.bio.ed.ac.uk/software/tracer.
- **FigTree** – this is an application for displaying and printing molecular phylogenies, in particular those obtained using BEAST. At the time of writing, the current version is v1.4.2. You can download FigTree from http://tree.bio.ed.ac.uk/software/figtree.

This chapter will guide you through the analysis of an alignment of sequences sampled from 12 primate species (see Figure 1.2). The goal is to estimate the phylogeny, the rate of evolution on each lineage and the ages of the uncalibrated ancestral divergences.

The first step will be to convert a NEXUS file with a DATA or CHARACTERS block into a BEAST XML input file. This is done using the program BEAUti (which stands for Bayesian Evolutionary Analysis Utility). This is a user-friendly program for setting the evolutionary model and options for the MCMC analysis. The second step is to run BEAST using the input file generated by BEAUti, which contains the data, model and analysis settings. The final step is to explore the output of BEAST in order to diagnose problems and to summarise the results.

6.1   BEAUti

The program BEAUti is a user-friendly program for setting the model parameters for BEAST. Run BEAUti by double-clicking on its icon. Once running, `BEAUti` will look similar irrespective of which computer system it is running on. For this chapter, the Mac OS X version is used in the figures but the Linux and Windows versions will have the same layout and functionality.

6.1.1   Loading the NEXUS file

To load a NEXUS format alignment, simply select the `Import Alignment...` option from the File menu, or drag the file into the middle of the Partitions panel.

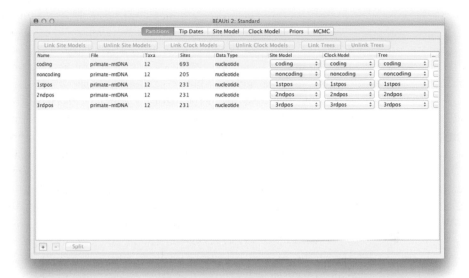

Figure 6.1 A screenshot of the data tab in BEAUti. This and all following screenshots were taken on an Apple computer running Mac OS X and will look slightly different on other operating systems.

The example file called `primates-mtDNA.nex` is available from the `examples/nexus` directory for Mac and Linux and `examples\nexus` for Windows inside the directory where BEAST is installed. This file contains an alignment of sequences of 12 species of primates.

Once loaded, five character partitions are displayed in the main panel (Figure 6.1). The alignment is divided into a protein coding part and a non-coding part, and the coding part is divided in codon positions 1, 2 and 3. You must remove the 'coding' partition before continuing to the next step as it refers to the same nucleotides as partitions '1stpos', '2ndpos' and '3rdpos'. To remove the 'coding' partition select the row and click the '-' button at the bottom of the table. You can view the alignment by double-clicking the partition.

Link/unlink partition models

Since the sequences are linked (i.e. they are all from the mitochondrial genome, which is not believed to undergo recombination in birds and mammals) they share the same ancestry, so the partitions should share the same time-tree in the model. For the sake of simplicity, we will also assume the partitions share the same evolutionary rate for each branch, and hence the same 'clock model'. We will restrict our modelling of rate heterogeneity to among-site heterogeneity within each partition, and also allow the partitions to have different mean rates of evolution. At this point we will need to link the clock model and tree. In the **Partitions** panel, select all four partitions in the table (or none, by default all partitions are affected) and click the `Link Tree Models` button and then the `Link Clock Models` button (see Figure 6.2). Then click on the

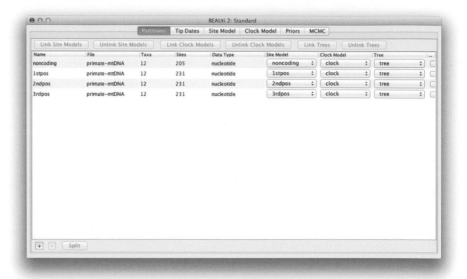

Figure 6.2 A screenshot of the **Partitions** tab in BEAUti after linking and renaming the clock model and tree.

first drop-down menu in the Clock Model column and rename the shared clock model to 'clock'. Likewise, rename the shared tree to 'tree'. This will make following options and generated log files more easy to read.

Setting the substitution model

The next step is to set up the substitution model. Select the **Site Model** tab at the top of the main window (we skip the Tip Dates tab since all taxa are from contemporary samples). This will reveal the evolutionary model settings for BEAST. The options available depend on whether the data are nucleotides, amino acids, binary data or general data. The settings that will appear after loading the primate nucleotide alignment will be the default values for nucleotide data, so we need to make some changes.

Most of the models should be familiar to you (see Chapter 3 for details). First, set the **Gamma Category Count** to 4 and then tick the 'estimate' box for the **Shape** parameter. This will allow rate variation between sites in each partition to be modelled. Note that 4–6 categories works sufficiently well for most data sets, while having more categories takes more time to compute for little added benefit. We leave the Proportion Invariant entry set to zero.

Select **HKY** from the **Subst Model** drop-down menu (Figure 6.3). Ideally, a substitution model should be selected that fit the data best for each partition, but here for the sake of simplicity we use HKY for all partitions. Further, select **Empirical** from the **Frequencies** drop-down menu. This will fix the frequencies to the proportions observed in the data (for each partition individually, once we clone the model settings to the other partitions). This approach means that we can get a good fit to the data without explicitly estimating these parameters. We do it here simply to make the log files a bit shorter

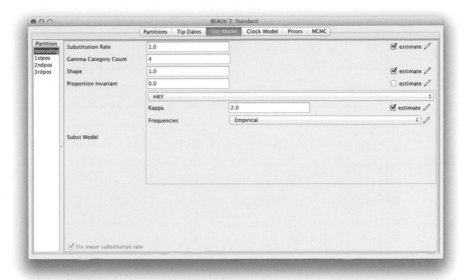

Figure 6.3 A screenshot of the **Site Model** tab in BEAUti.

and more readable in later parts of the exercise. Finally check the 'estimate' box for the **Substitution rate** parameter and also check the **Fix mean substitution rate** box. This will allow the individual partitions to have their relative rates estimated for the unlinked site models.

Last, hold 'shift' key to select all site models on the left side, and click **OK** to clone the setting from **noncoding** into **1stpos**, **2ndpos** and **3rdpos** (Figure 6.3). Go through each site model, as you can see, their configurations are the same now.

Setting the clock model

The next step is to select the **Clock Model** tab at the top of the main window. This is where we select the molecular clock model. For this exercise we are going to leave the selection at the *default* value of a strict molecular clock, because these data are very clock-like and do not need rate variation among branches to be included in the model. To test for clock-likeness, you can (1) run the analysis with a relaxed clock model and check how much variation among rates is implied by the data (see coefficient of variation in Chapter 10 for more on this), or (2) perform a model comparison between a strict and relaxed clock using path sampling, or (3) use a random local clock model (Drummond and Suchard 2010) which explicitly considers whether each branch in the tree needs its own branch rate.

Priors

The **Priors** tab allows priors to be specified for each parameter in the model. The model selections made in the Site Model and Clock Model tabs result in the inclusion of various parameters in the model, and these are shown in the Priors tab (see Figure 6.4).

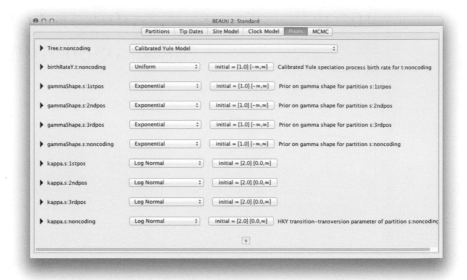

Figure 6.4 A screenshot of the **Priors** tab in BEAUti.

Here we also specify that we wish to use the Calibrated Yule model (Heled and Drummond 2012) as the tree prior. The Yule model is a simple model of speciation that is generally more appropriate when considering sequences from different species (see Section 2.4 for details). Select this from the **Tree prior** drop-down menu.

We should convince ourselves that the priors shown in the priors panel really reflect the prior information we have about the parameters of the model. We will specify diffuse 'uninformative' but proper priors on the overall molecular clock rate (`clockRate`) and the speciation rate (`birthRateY`) of the Yule tree prior. For each of these parameters select **Gamma** from the drop-down menu and, using the arrow button, expand the view to reveal the parameters of the gamma prior. For both the clock rate and the Yule birth rate set the Alpha (shape) parameter to 0.001 and the Beta (scale) parameter to 1000.

By default each of the gamma shape parameters has an exponential prior distribution with a mean of 1. This implies (see Figure 3.7) we expect some rate variation. By default the kappa parameters for the HKY model have a log-normal (1,1.25) prior distribution, which broadly agrees with empirical evidence (Rosenberg et al. 2003) on the range of realistic values for transition/transversion bias. These default priors are kept since they are suitable for this particular analysis.

6.1.2   Defining the calibration node

We now need to specify a prior distribution on the calibrated node, based on our fossil knowledge. This is known as calibrating our tree. To define an extra prior on the tree, press the small **+** button below list of priors. You will see a dialogue that allows you to define a subset of the taxa in the phylogenetic tree. Once you have created a taxa set you will be able to add calibration information for its most recent common ancestor (MRCA) later on.

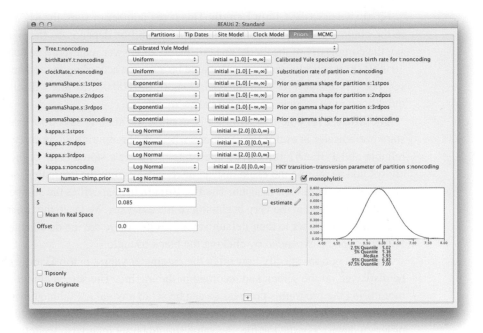

Figure 6.5 A screenshot of the calibration prior options in the Priors panel in BEAUti.

Name the taxa set by filling in the taxon set label entry. Call it human-chimp, since it will contain the taxa for *Homo sapiens* and *Pan*. In the list below you will see the available taxa. Select each of the two taxa in turn and press the >> arrow button. Click OK and the newly defined taxa set will be added to the prior list. As this is a calibrated node to be used in conjunction with the Calibrated Yule prior, monophyly must be enforced, so select the checkbox marked Monophyletic?. This will constrain the tree topology so that the human–chimp grouping is kept monophyletic during the course of the MCMC analysis.

To encode the calibration information we need to specify a distribution for the most recent common ancestor (MRCA) of human–chimp. Select the **Log-normal** distribution from the drop-down menu to the right of the newly added human-chimp.prior. Click on the black triangle and a graph of the probability density function will appear, along with parameters for the log-normal distribution. We are going to set $M = 1.78$ and $S = 0.085$, which will specify a distribution centred at about 6 million years with a standard deviation of about 0.5 million years. This will give a central 95% probability range covering 5–7 Mya. This roughly corresponds to the current consensus estimate of the date of the MRCA of humans and chimpanzees (Figure 6.5).

Setting the MCMC options

The next tab, **MCMC**, provides more general settings to control the length of the MCMC run and the file names.

Firstly we have the **Chain Length**. This is the number of steps the MCMC will make in the chain before finishing. How long this should be depends on the size of the data set, the complexity of the model and the quality of answer required. The default value of 10 000 000 is entirely arbitrary and should be adjusted according to the size of your data set. For this analysis let's set the chain length to 6 000 000 as this will run reasonably quickly on most modern computers (a few minutes).

The **Store Every** field determines how often the state is stored to file. Storing the state periodically is useful for situations where the computing environment is not very reliable and a BEAST run can be interrupted. Having a stored copy of the recent state allows you to resume the chain instead of restarting from the beginning, so you do not need to get through burn-in again. The **Pre Burnin** field specifies the number of samples that are not logged at the very beginning of the analysis. We leave the **Store Every** and **Pre Burnin** fields set to their default values. Below these are the details of the log files. Each one can be expanded by clicking the black triangle.

The next options specify how often the parameter values in the Markov chain should be displayed on the screen and recorded in the log file. The screen output is simply for monitoring the program's progress so can be set to any value (although if set too small, the sheer quantity of information being displayed on the screen will actually slow the program down). For the log file, the value should be set relative to the total length of the chain. Sampling too often will result in very large files with little extra benefit in terms of the accuracy of the analysis. Sample too infrequently and the log file will not record sufficient information about the distributions of the parameters. You probably want to aim to store no more than 10 000 samples, so this should be set to no less than chain length / 10 000.

For this exercise we will set the screen log to 10 000 and the file log to 1000. The final two options give the file names of the log files for the sampled parameters and the trees. These will be set to a default based on the name of the imported NEXUS file.

If you are using the Windows operating system then we suggest you add the suffix `.txt` to both of these (so, `Primates.log.txt` and `Primates.trees.txt`) so that Windows recognises these as text files.

Generating the BEAST XML file

We are now ready to create the BEAST XML file. To do this, select the **Save** option from the **File** menu. Check the default priors, and save the file with an appropriate name (we usually end the filename with `.xml`, i.e. `Primates.xml`). We are now ready to run the file through BEAST.

6.2   Running BEAST

Now run BEAST and when it asks for an input file (Figure 6.6), provide your newly created XML file as input. BEAST will then run until it has finished reporting information to the screen. The actual results files are saved to the disk in the same location as your input file. The output to the screen will look something like this:

```
                      BEAST v2.2.1, 2002-2015
              Bayesian Evolutionary Analysis Sampling Trees
                      Designed and developed by
      Remco Bouckaert, Alexei J. Drummond, Andrew Rambaut & Marc A. Suchard

                      Department of Computer Science
                         University of Auckland
                         remco@cs.auckland.ac.nz
                         alexei@cs.auckland.ac.nz

                      Institute of Evolutionary Biology
                         University of Edinburgh
                           a.rambaut@ed.ac.uk

                      David Geffen School of Medicine
                    University of California, Los Angeles
                           msuchard@ucla.edu

                      Downloads, Help & Resources:
                           http://beast2.org/

         Source code distributed under the GNU Lesser General Public License:
                      http://github.com/CompEvol/beast2

                           BEAST developers:
           Alex Alekseyenko, Trevor Bedford, Erik Bloomquist, Joseph Heled,
         Sebastian Hoehna, Denise Kuelnert, Philippe Lemey, Wai Lok Sibon Li,
         Gerton Lunter, Sidney Markowitz, Vladimir Minin, Michael Defoin Platel,
                    Oliver Pybus, Chieh-Hsi Wu, Walter Xie

                                Thanks to:
                Roald Forsberg, Beth Shapiro and Korbinian Strimmer

      Random number seed: 777

      Alignment(primate-mtDNA)
        12 taxa
        898 sites
        413 patterns
```

Figure 6.6 A screenshot of BEAST.

```
TreeLikelihood(treeLikelihood.noncoding) uses BeerLikelihoodCore4
  FilteredAlignment(noncoding): [taxa, patterns, sites] = [12, 79, 205]
TreeLikelihood(treeLikelihood.1stpos) uses BeerLikelihoodCore4
  FilteredAlignment(1stpos): [taxa, patterns, sites] = [12, 120, 231]
TreeLikelihood(treeLikelihood.2ndpos) uses BeerLikelihoodCore4
  FilteredAlignment(2ndpos): [taxa, patterns, sites] = [12, 82, 231]
TreeLikelihood(treeLikelihood.3rdpos) uses BeerLikelihoodCore4
  FilteredAlignment(3rdpos): [taxa, patterns, sites] = [12, 204, 231]
=====================================================================
Citations for this model:

Bouckaert RR, Heled J, Kuehnert D, Vaughan TG, Wu C-H, Xie D, Suchard MA,
  Rambaut A, Drummond AJ (2014) BEAST 2: A software platform for Bayesian
  evolutionary analysis. PLoS Computational Biology 10(4): e1003537

Heled J, Drummond AJ (2012) Calibrated Tree Priors for Relaxed Phylogenetics
  and Divergence Time Estimation. Systematic Biology 61(1):138-149.

Hasegawa M, Kishino H, Yano T (1985) Dating the human-ape splitting by a
  molecular clock of mitochondrial DNA. Journal of Molecular Evolution
  22:160-174.

=====================================================================
Writing file Primates.log
Writing file Primates.trees
      Sample      posterior ESS(posterior)     likelihood        prior
           0    -7924.3599              N     -7688.4922   -235.8676 --
       10000    -5529.0700            2.0     -5459.1993    -69.8706 --
       20000    -5516.8159            3.0     -5442.3372    -74.4786 --
       30000    -5516.4959            4.0     -5439.0839    -77.4119 --
       40000    -5521.1160            5.0     -5445.6047    -75.5113 --
       50000    -5520.7350            6.0     -5444.6198    -76.1151 --
       60000    -5512.9427            7.0     -5439.2561    -73.6866 2m31s/Msamples
       70000    -5513.8357            8.0     -5437.9432    -75.8924 2m31s/Msamples

         ...           ...            ...            ...          ...
     5990000    -5516.6832          474.6     -5442.5945    -74.0886 2m31s/Msamples
     6000000    -5512.3802          472.2     -5440.8928    -71.4874 2m31s/Msamples
```

Operator	Tuning	#accept	#reject	Pr(m)	Pr(acc|m)
ScaleOperator(treeScaler.t:tree)	0.7032	39935	174155	0.0358	0.1865
ScaleOperator(treeRootScaler.t:tree)	0.6438	37329	177166	0.0358	0.1740
Uniform(UniformOperator.t:tree)	-	479419	1668915	0.3580	0.2232
SubtreeSlide(SubtreeSlide.t:tree)	9.9217	272787	801404	0.1790	0.2539
Exchange(narrow.t:tree)	-	744	1074261	0.1790	0.0007
Exchange(wide.t:tree)	-	9	214594	0.0358	0.0000
WilsonBalding(WilsonBalding.t:tree)	-	4	214548	0.0358	0.0000
ScaleOperator(KappaScaler.s:noncoding)	0.3521	1739	5375	0.0012	0.2444
DeltaExchangeOperator(FixMeanMutationRatesOperator)	0.4252	17277	126203	0.0239	0.1204
ScaleOperator(gammaShapeScaler.s:noncoding)	0.3749	1729	5428	0.0012	0.2416
ScaleOperator(CalibratedYuleBirthRateScaler.t:tree)	0.2447	58005	156128	0.0358	0.2709
ScaleOperator(StrictClockRateScaler.c:clock)	0.7064	50080	164952	0.0358	0.2329
UpDownOperator(strictClockUpDownOperator.c:clock)	0.5888	50809	163882	0.0358	0.2367
ScaleOperator(KappaScaler.s:1stpos)	0.4398	1816	5388	0.0012	0.2521
ScaleOperator(gammaShapeScaler.s:1stpos)	0.4203	1927	5129	0.0012	0.2731
ScaleOperator(KappaScaler.s:2ndpos)	0.3321	1964	5301	0.0012	0.2703
ScaleOperator(gammaShapeScaler.s:2ndpos)	0.3029	2033	5177	0.0012	0.2820
ScaleOperator(KappaScaler.s:3rdpos)	0.5052	1424	5860	0.0012	0.1955
ScaleOperator(gammaShapeScaler.s:3rdpos)	0.2670	1569	5536	0.0012	0.2208

```
    Tuning: The value of the operator's tuning parameter, or '-' if the operator can't be optimized.
   #accept: The total number of times a proposal by this operator has been accepted.
   #reject: The total number of times a proposal by this operator has been rejected.
    Pr(m): The probability this operator is chosen in a step of the MCMC (i.e. the normalized weight).
 Pr(acc|m): The acceptance probability (#accept as a fraction of the total proposals for this operator).

Total calculation time: 910.661 seconds
```

Note that there is some useful information at the start concerning the alignments and which tree likelihoods are used. Also, all citations relevant for the analysis are mentioned at the start of the run, which can easily be copied to manuscripts reporting about the analysis. Then follows reporting of the chain, which gives some real-time feedback on progress of the chain.

At the end, an operator analysis is printed, which lists all operators used in the analysis together with how often the operator was tried, accepted and rejected (see columns #total, #accept and #reject, respectively). The acceptance rate is the proportion of times an operator is accepted when it is selected for doing a proposal. In general, an acceptance rate that is high, say over 0.5, indicates the proposals are conservative and do not explore the parameter space efficiently. On the other hand, a low acceptance rate indicates that proposals are too aggressive and almost always result in a state that is rejected because of its low posterior. Both too high and too low acceptance rates result in low effective sample size (ESS) values. An acceptance rate of 0.234 is the target (based on very limited evidence provided by Gelman et al. 1996) for many (but not all) operators implemented in BEAST.

Some operators have a tuning parameter, for example the scale factor of a scale parameter. If the final acceptance rate is not near the target, BEAST will suggest a new value for the tuning parameter, which is printed in the operator analysis. In this case, all acceptance rates are good for the operators that have tuning parameters. Operators without tuning parameters include the wide exchange and Wilson–Balding operators for this analysis. Both of these operators attempt to change the topology of the tree with large steps, but since the data support a single topology overwhelmingly, these radical proposals are almost always rejected.

6.3 Analysing the results

The BEAST run produces two logs; a trace log and a tree log. To inspect the trace log, run the program called **Tracer**. When the main window has opened, choose **Import Trace File...** from the **File** menu and select the file that BEAST has created called `Primates.log` (Figure 6.7).

Remember that MCMC is a stochastic algorithm so the actual numbers will not be exactly the same as those depicted in the figure.

On the left-hand side is a list of the different quantities that BEAST has logged to file. There are traces for the posterior (this is the natural logarithm of the product of the tree likelihood and the prior density), and the continuous parameters. Selecting a trace on the left brings up analyses for this trace on the right-hand side, depending on the tab that is selected. When first opened, the 'posterior' trace is selected and various statistics of this trace are shown under the Estimates tab. In the top right of the window is a table of calculated statistics for the selected trace.

Select the `clockRate` parameter in the left-hand list to look at the average rate of evolution (averaged over the whole tree and all sites). Tracer will plot a (marginal posterior) histogram for the selected statistic and also give you summary statistics such as the mean and median. The 95% HPD interval stands for *highest posterior density* interval and represents the most compact interval on the selected parameter that contains 95% of the posterior probability. It can be loosely thought of as a Bayesian analogue to a confidence interval. The `TreeHeight` parameter gives the marginal posterior distribution of the age of the root of the entire tree.

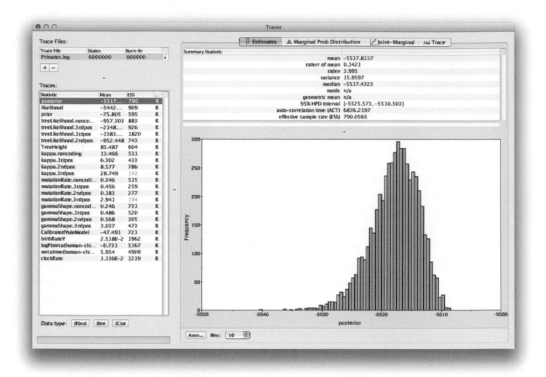

Figure 6.7 A screenshot of Tracer v1.6.

Select the `TreeHeight` parameter and then Ctrl-click `mrcatime (human-chimp)` (Command-click on Mac OS X). This will show a display of the age of the root and the calibration MRCA we specified earlier in BEAUti. You can verify that the divergence that we used to calibrate the tree (`mrcatime(human-chimp)`) has a posterior distribution that matches the prior distribution we specified (Figure 6.8).

Here are a few questions to consider in light of the Tracer summary:

- What is the estimated rate of molecular evolution for this gene tree (include the 95% HPD interval)?
- What sources of error does this estimate include?
- How old is the root of the tree (give the mean and the 95% HPD range)?

6.4   Marginal posterior estimates

To show the relative rates for the four partitions, select the mutationRate parameter for each of the four partitions, and select the marginal density tab in Tracer. Figure 6.9 shows the marginal densities for the relative substitution rates. The plot shows that codon positions 1 and 2 have substantially different rates (0.452 versus 0.181) and both are far slower than codon position 3 with a relative rate of 2.95. The non-coding partition

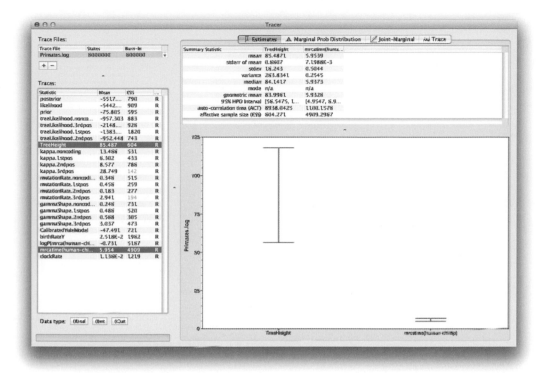

Figure 6.8 A screenshot of the 95% HPD intervals of the root height and the user-specified (human–chimp) MRCA in Tracer.

has a rate intermediate between codon positions 1 and 2 (0.344). Taken together this result suggests strong purifying selection in both the coding and non-coding regions of the alignment.

Likewise, a marginal posterior estimate can be obtained for the gamma shape parameter and the kappa parameter, which are shown in Figures 6.10 and 6.11, respectively. The plot for the gamma shape parameter suggest that there is considerable rate variation for all of the partitions with the least rate variation in the third codon position.

The plot for the kappa parameter (Figure 6.11) shows that all partitions show considerable transition/transversion bias, but that the third codon position in particular has a high bias with a mean of almost 29 more transitions than transversions.

6.5   Obtaining an estimate of the phylogenetic tree

BEAST also produces a posterior sample of phylogenetic time-trees along with its sample of parameter estimates. These can be summarised using the program **TreeAnnotator**. This will take the set of trees and find the best-supported one. It will then annotate this representative summary tree with the mean ages of all the nodes and the corresponding 95% HPD ranges. It will also calculate the posterior clade

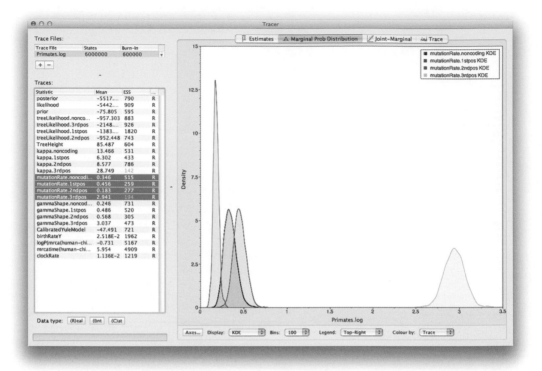

Figure 6.9 A screenshot of the marginal posterior densities of the relative substitution rates of the four partitions (relative to the site-weighted mean rate).

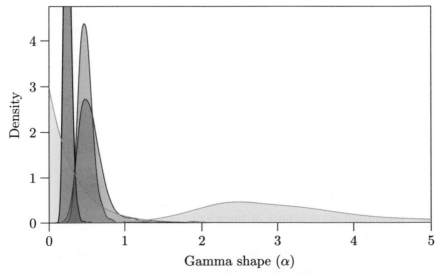

Figure 6.10 The marginal prior and posterior densities for the shape (α) parameters. The prior is in green. The posterior density estimate for each partition is also shown: non-coding (black) and first (blue), second (red) and third (orange) codon positions.

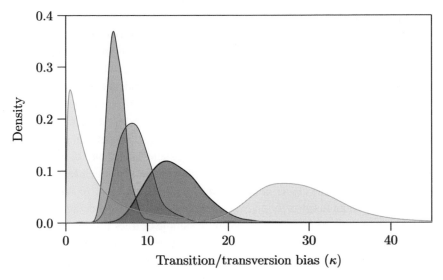

Figure 6.11 The marginal prior and posterior densities for the transition/tranversion bias (κ) parameters. The prior is in green. The posterior density estimate for each partition is also shown: non-coding (black) and first (blue), second (red) and third (orange) codon positions.

Figure 6.12 A screenshot of TreeAnnotator.

probability for each node. Run the TreeAnnotator program and set it up as depicted in Figure 6.12.

The burn-in is the number of trees to remove from the start of the sample. Unlike **Tracer**, which specifies the number of steps as a burn-in, in **TreeAnnotator** you need to specify the actual number of trees. For this run, you specified a chain length of 6 000 000 steps, sampling every 1000 steps. Thus the trees file will contain 5000 trees and so to specify a 1% burn-in use the value 50.

The **Posterior probability limit** option specifies a limit such that if a node is found at less than this frequency in the sample of trees (i.e. has a posterior probability less than this limit), it will not be annotated. The default of 0.5 means that only nodes seen in the majority of trees will be annotated. Set this to zero to annotate all nodes.

The **Target tree type** specifies the tree topology that will be annotated. You can either choose a specific tree from a file or ask TreeAnnotator to find a tree in your sample. The default option, **Maximum clade credibility tree**, finds the tree with the highest product of the posterior probability of all its nodes.

For node heights, the default is Common Ancestor Heights, which calculates the height of a node as the mean of the MRCA time of all pairs of nodes in the clade. For trees with large uncertainty in the topology and thus many clades with low support, some other methods can result in trees with negative branch lengths. In this analysis, the support for all clades in the summary tree is very high, so this is not an issue here. Choose **Mean heights** for node heights. This sets the heights (ages) of each node in the tree to the mean height across the entire sample of trees for that clade.

For the input file, select the trees file that BEAST created and select a file for the output (here we called it `Primates.MCC.tree`). Now press Run and wait for the program to finish.

6.6   Visualising the tree estimate

Finally, we can visualise the tree in another program called **FigTree**. Run this program, and open the `Primates.MCC.tree` file by using the Open command in the File menu. The tree should appear. You can now try selecting some of the options in the control panel on the left. Try selecting **Node Bars** to get node age error bars. Also turn on **Branch Labels** and select **posterior** to get it to display the posterior probability for each node. Under **Appearance** you can also tell FigTree to colour the branches by the rate. You should end up with something similar to Figure 6.13.

An alternative view of the tree can be made with DensiTree, which is part of BEAST 2. The advantage of DensiTree is that it is able to visualise both uncertainty in node heights and uncertainty in topology. For this particular data set, the most probable topology is present in more than 99% of the samples. So, we conclude that this analysis results in a very high consensus on topology (Figure 6.13).

6.7   Comparing your results to the prior

It is a good idea to rerun the analysis while sampling from the prior to make sure that interactions between priors are not affecting your prior information. The interaction between priors can be problematic, especially when using calibrations since it means putting multiple priors on the tree (see Section 9.1 for more details). Using BEAUti, you can set up the same analysis under the MCMC options by selecting the **Sample from prior only** option. This will allow you to visualise the full prior distribution in the absence of your sequence data. Summarise the trees from the full prior distribution and compare the summary to the posterior summary tree.

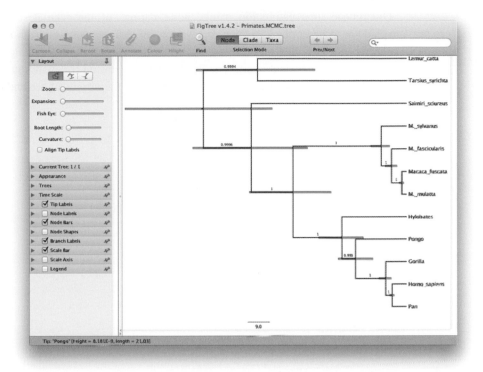

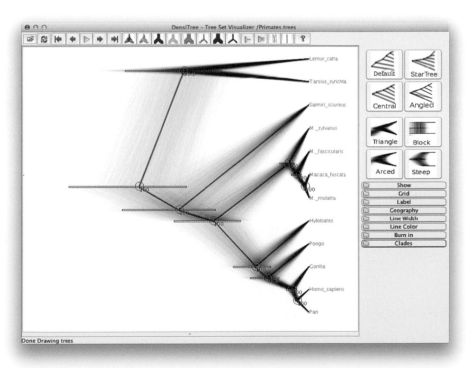

Figure 6.13 A screenshot of FigTree (top) and DensiTree (bottom) for the primate data.

Divergence time estimation using 'node dating' of the type described in this chapter has been applied to answer a variety of different questions in ecology and evolution. For example, node dating with fossils was used in determining the species diversity of cycads (Nagalingum et al. 2011), analysing the rate of evolution in flowering plants (Smith and Donoghue 2008) and investigating the origins of hot and cold desert cyanobacteria (Bahl et al. 2011).

7   Setting up and running a phylogenetic analysis

In this chapter, we will go through some of the more common decisions involved in setting up a phylogenetic analysis in BEAST. The order in which the issues are presented follows more or less the order in which an analysis is set up in BEAUti for a standard analysis. So, we start with issues involved in the alignment, then setting up site and substitution models, clock models and tree priors and all of their priors. Some notes on calibrations and miscellanea are followed by some practicalities of running a BEAST analysis. Note that a lot of the advice in this section is rather general. Since every situation has its special characteristics, the advice should be interpreted in the context of what you know about your data.

7.1   Preparing alignments

Some tips on selecting samples and loci for alignments are discussed in (Ho and Shapiro 2011; Mourier et al. 2012; Silva et al. 2012).

Recombinant sequences: Though under some circumstances, horizontal transmission was shown not to impact the tree and divergence time estimates (Currie et al. 2010; Greenhill et al. 2010), the models in BEAST cannot handle recombinant sequences properly at the time of writing. So, it is recommended that these are removed from the alignment. There are many programs that can help identify recombinant sequences, for example 3seq (Boni et al. 2007) or SplitsTree (Huson and Bryant 2006).

Duplicate sequences: An erroneous argument for removal of duplicate sequences in the alignment is that multiple copies will lead to ambiguous trees and slow down the analysis. However, a Bayesian approach aims to sample all trees that have an appreciable probability given the data. One of the assumptions underlying common Bayesian phylogenetic models is that there is a binary tree according to which the data were generated. If, for example, three taxa have identical sequences, it does not mean that they represent the same individual, or that they are equally closely related in the true tree. All that can be said is that there were no mutations in the sampled part of the genome during the ancestral history of those three taxa. In this case, BEAST would sample all three subtrees with equal probability: ((1,2),3), (1,(2,3)), ((1,3),2). If you summarise the BEAST output as a single tree (see Section 11.4) you will see some particular sub-tree over these identical sequences based on the selected representative tree. But the posterior probability for that particular sub-tree will probably be low (around one-third in our example), since other trees have also been sampled in the chain.

One of the results of a Bayesian phylogenetic analysis is that it gives an estimate of how closely related the sampled sequences are, even if the sequences are identical. This is possible because all divergences in the phylogeny are estimated using a probabilistic model of substitution. For identical sequences, this amounts to determining how old the common ancestor of these sequences could be given that no mutations were observed in their common ancestry, and given the estimated substitution rate and sequence length. Among a set of identical sequences, the only divergence with the possibility of significant support would be their common ancestor. If this is the case then you can confidently report the age of their common ancestor, but should not try to make any statements about relationships or divergence times within the group of identical sequences.

Finally, there is a population genetic reason not to remove identical sequences. Imagine you have sequenced 100 random individuals and among them you observe only 20 unique haplotypes. You are tempted to just analyse the 20 haplotypes. However, if you are applying a population genetic prior like the coalescent, then this is equivalent to misrepresenting the data, since the coalescent tree prior assumes that you have randomly sampled the 20 individuals. If only unique haplotypes are analysed, then it will appear that every random individual sampled had a unique haplotype. If this was actually the case you would conclude that the background population from which these individuals came must be very large. As a result, by removing all the identical sequences you will cause an overestimation of the population size.

Outgroups: An outgroup is a taxon or set of taxa that is closely related to the taxa of interest (the ingroup), but definitely has a common ancestor with the ingroup that is more ancient then the most recent common ancestor of the ingroup. In unrooted phylogenetic reconstruction the outgroup traditionally serves as a phylogenetic reference and provides a root for the ingroup (Felsenstein 2004). However, adding an outgroup is generally discouraged in Bayesian time-tree analyses because inclusion of outgroups can introduce long branches which can make many estimation tasks more difficult. Having said that, a well-chosen outgroup can provide additional information about the ingroup root position, even when a molecular clock is already being used to estimate a rooted tree.

Nevetheless, outgroups are usually less well sampled than the ingroup, which violates a basic assumption of many of the standard time-tree priors. Most time-tree priors assume that the entire tree is sampled with consistent intensity across all clades at each sampling time. For example, this assumption underlies most coalescent and birth–death priors (but for alternative sampling assumptions see Höhna et al. 2011).

Also, in a population genetic context, if the outgroup is from a different species than the ingroup and the ingroup taxa are from the same species, care should be taken in selecting a prior, and your options may be limited, compared to analyses restricted to the single-species ingroup. Coalescent-based priors are appropriate for the ingroup, but birth–death-based priors are more appropriate for divergences separating different species in the tree.

Table 7.1 IUPAC-IUB ambiguity codes for nucleotide data

Code	Ambiguities	Code	Ambiguities
R	A, G	B	C, G, T
Y	C, T	D	A, G, T
M	A, C	H	A, C, T
W	A, T	V	A, C, G
S	C, G	N,?,-	A, C, G, T
K	G, T		

Finally, traditionally the outgroup was picked to be the one most genetically similar to the ingroup, that is, on the shortest branch. However, this tends to select for atypical taxa that are evolving slowly, which has a biasing impact on the relaxed and strict clock analyses required to do divergence-time dating (Suchard et al. 2003).

So, the simple answer to the question 'How do I instruct BEAST to use an outgroup?' is that you may not want to. A Bayesian time-tree analysis will sample the root position along with the rest of the tree topology. If you then calculate the proportion of sampled trees that have a particular root, you obtain a posterior probability for the root position. However, if you do include an outgroup and have a strong prior belief that it really is an outgroup then you should probably reflect that in your model by constraining the ingroup to be monophyletic.

Ambiguous data: When a site in a sequence cannot be unambiguously determined, but it is known to be from a subset of characters, these positions in the sequence can be encoded as ambiguous. For example, for nucleotide data an 'R' in a sequence represents the state is either 'A' or 'G' but certainly not 'C' or 'T' (see Table 7.1 for the IUPAC-IUB ambiguity codes for nucleotide data). By default, ambiguous data are treated as unknowns for reasons of efficiency, so internally they are replaced by a '?' or '-'. Both unknowns and gaps in the sequence are treated the same in most likelihood-based phylogenetic analyses. Note there are alternative approaches that use indels as phylogenetic information (Lunter et al. 2005; Novák et al. 2008; Redelings and Suchard 2005, 2007; Suchard and Redelings 2006). By treating ambiguities as unknowns the phylogenetic likelihood (see Section 3.5) can be calculated about twice as fast as when ambiguities are explicitly modelled. When a large proportion of the data consists of ambiguities, it may be worth modelling the ambiguities exactly. This can be done by setting the useAmbiguities="true" flag on the tree-likelihood element in the XML (see Chapter 13).

However, simulations and empirical analysis suggest that missing data are not problematic and for sufficiently long sequences taxa with missing data will be accurately placed in the phylogeny. When the number of characters in the analysis is larger than 100, and up to 95% of the sequence data are missing the tree can still be reconstructed correctly (Wiens and Moen 2008). Model misspecification is probably a larger problem in assuring phylogenetic accuracy than missing data (Roure et al. 2013).

Partitioning: Alignments can be split into various subsets called partitions. For example, if the alignment is known to be protein encoding, it often makes sense to split according to codon position (Bofkin and Goldman 2007; Shapiro et al. 2006). If there is a coding and non-coding part of a sequence, it may make sense to separate the two parts in different partitions (see for instance the exercise in Chapter 6). Partitions can be combined by linking site models, clock models and trees for the partitions of interest. This has the effect of appending the alignments in the partitions. Which of the various combinations of linking and unlinking site models, clock models and trees is appropriate depends on the scenario. For example:

Scenario 1: If you are interested in the gene tree from multiple genes sampled from a single evolutionarily linked molecule like the mitochondrial genome or the hepatitis C virus (HCV) genome, then it is more likely you are interested in estimating a single phylogenetic tree from a number of different genes, and for each gene you would like a different substitution model and relative rate.

Scenario 2: If you are interested in a species tree from multiple genes that are not linked, then a multispecies coalescent (*BEAST; Heled and Drummond 2010) analysis is more appropriate.

Scenario 3: If you are interested in estimating the average birth rate of lineages from a number of different phylogenetic data sets you would set up a multi-partition analysis. But you would have a series of phylogenetic trees that all share the same Yule birth rate parameter to describe the distribution of branch lengths (instead of a series of population genealogies sharing a population size parameter). In addition they may have different relative rates of evolution and different substitution models, which you could set up by having multiple site models, one for each partition.

If one gene sequence is missing in one taxa then it should be fine to just use '?' to represent this (or gaps, '-', they are treated the same way). If the shorter sequences can be assumed to follow a different evolutionary path than the remainder, it is better to split the sequences into partitions that share a single tree, but allow different substitution models.

Partitions should not be chosen too small. There should still be sufficient information in each partition to be able to estimate parameters of interest. One has to keep in mind that it is not the number of sites in the partition that matters, but the number of unique site patterns. A site pattern is an assignment of characters in an alignment for a particular site, illustrated as any column in the alignment in Figure 1.2.

Methods that automatically pick partitions from data include (Wu et al. 2013) for arbitrary partitions and (Bouckaert et al. 2013) for consecutive partitions.

Combining partitions: You can combine partitions if they share the same taxa simply by linking their tree, site model and clock model. This has the same effect as concatenating the alignments into a single partition. Note that sharing just part of the model across multiple partitions is also possible.

7.2   Choosing priors/model set-up

In general, it is very hard to give an answer to the question 'Which prior should I use?', because the only proper answer is 'It depends'. In this section we make an attempt to

Table 7.2 Some common distributions used as priors and some of their properties. Plots indicate some of the shapes

Probability density function		Parameter	Effect of increasing parameter on distribution	Range*
Normal $N(x\|\mu,\sigma) =$ $\frac{1}{\sqrt{2\pi\sigma^2}}e^{-(x-\mu)^2/\sigma^2}$		Mean μ Standard deviation σ	Shift to right Make distribution wider and flatter	$(-\infty, \infty)$
Log-normal $LN(x\|M,S) =$ $\frac{1}{x\sqrt{2\pi S^2}}e^{-(\ln(x)-M)^2/S^2}$		Mean $M^{\dagger}$ Standard deviation S Offset o	Shift to right Make distribution wider and flatter Shift to right	$[0, \infty)$
Gamma $\Gamma(x\|\alpha,\beta)^{\ddagger} =$ $\frac{1}{\Gamma(\alpha)\beta^k}x^{\alpha-1}e^{-x/\beta}$		Shape α Scale β Offset o	Concentrate into peak at $(\alpha\beta)$ once $\alpha > 1$ Flattens Shift to right	$[0, \infty)$
Inverse gamma $I\Gamma(x\|\alpha,\beta) =$ $\frac{\beta^{\alpha}}{\Gamma(\alpha)}x^{-\alpha-1}e^{-\beta/x}$		Shape α Scale β Offset o	Concentrate into peak Flattens Shift to right	$[0, \infty)$
Beta $Beta(x\|\alpha,\beta) =$ $\frac{\Gamma(\alpha+\beta)}{\Gamma(\alpha)\Gamma(\beta)}x^{\alpha-1}$ $(1.0 - x)^{\beta-1}$		Shape α Shape β Offset o	Shift mode left and concentrate Shift mode right and concentrate Shift to right	$[0, 1]$
Exponential $exp(x\|\lambda) =$ $\frac{1}{\lambda}e^{-x/\lambda}$		Scale λ Offset o	Increase mean and std.dev. linearly with λ Shift to right	$[0, \infty]$
Laplace $L(x\|\mu,b) =$ $\frac{1}{2b}e^{-\|x-m\|/b}$		Mean μ Scale b	Shift to right Increase std.dev. linearly with b	$(-\infty, \infty)$
1/X $OneOnX(x)$ $= \frac{1}{x}$		Offset o	Shift to right	$[0, \infty)$
uniform $U(x\|l,u) =$ $\begin{cases} \frac{1}{u-l} & \text{if } l \leq x \leq u \\ 0 & \text{otherwise} \end{cases}$		Lower l Upper u	Shift to left Shift to right	$(-\infty, \infty)$

* If offset is set to non-zero, the offset should be added to the range. † M is the mean of log x, but the log-normal distribution can also be specified by its true mean, μ. If so, μ is the mean of the distribution. ‡ NB a number of parameterisations are in use, this shows the one we use in BEAST.

give insight into the main issues around choosing priors. However, it is important to realise that for each of the pieces of advice given in the following paragraphs there are exceptions. As a reference, Table 7.2 shows some of the more common distributions used as priors.

7.2.1   Substitution and site models

One would like to choose a substitution model with as few parameters as possible, while capturing the evolutionary process as accurately as possible. These two competing requirements imply a few settings should be tried, where for nucleotide models HKY is a good starting point and after that try GTR if the HKY analysis converges satisfactorily. GTR can result in convergence issues in combination with complex models such as BSP and relaxed clock if all rates are estimated and the sequence alignment does not have enough variation/information in it. If there are few site patterns (see Section 3.7) with a particular combination of characters there may not be enough data to accurately estimate some of the rates.

Prior on HKY kappa parameter: Rosenberg et al. (2003) established an empirical distribution of κ from thousands of mammalian gene analyses. It is clear from this study that with long enough sequences and inter-species comparisons you end up with a κ between 1 and 10. But see (Yang and Yoder 1999) for the relationship between taxon sampling and κ estimation. So, in general a prior can be used that is generous on the upper side as long as the median is still within the 1–10 range. For example a log-normal with $M = 1$ and $S = 1.5$ which has a 95% range of about [0.14,51] and a median of 2.718 is probably suitable for most applications. However, note that for within-species comparisons κ can be high, with estimated κ values sometimes exceeding 40 (Yang and Yoder 1999). For example, in a survey of 5140 human mitochondrial genomes an average transition/transversion ratio (R) of 21.1:1 was found for polymorphisms over the 0.1% threshold (Pereira et al. 2009). Because these sequences are very closely related we can ignore multiple hits without too much compromise and the κ parameter can be estimated from $R = 21.1$ by the following simple conversion:

$$\kappa = \frac{R(\pi_A + \pi_G)(\pi_C + \pi_T)}{(\pi_A \pi_G + \pi_C \pi_T)}. \tag{7.1}$$

Assuming base frequencies of $\{\pi_A, \pi_C, \pi_G, \pi_T\} = \{0.309, 0.313, 0.131, 0.247\}$ (corresponding to those of the reference human mitochondrial genomes, accessions NC_012920 and J01415) gives a κ of 44.14 by this method.

General time-reversible (GTR) substitution model relative rates: The relative rates of the GTR model can be fixed when there is an external estimate available, but this is double-dipping if the values were calculated in another software package from the same data. Note that relative rates of the GTR model are normalised in different ways by different software packages so you cannot necessarily directly compare the absolute values of these relative rates. You should first make sure both sets of rates are normalised in the same way. For example many maximum likelihood software packages normalise so that the rate of G–T is 1; however in BEAST it is normally the rate of C–T changes

Table 7.3 GTR model can be used to define other nucleotide substitution models by linking parameters.

Model	α	β	γ	δ	ϵ	ω	# Dimensions
F81 (JC69)	1	1	1	1	1	1	0
HKY85 (K80)	a	1	a	a	1	a	1
TN93	a	b	a	a	1	a	2
TIM	a	b	c	c	1	a	3
GTR (SYM)	a	b	c	d	1	e	5

On the left, the four letters and the rates between them are shown. On the right, the models with frequencies estimated (models that have their frequencies not estimated in braces) and the corresponding values, which show which parameters are linked by sharing the same alphabetic character. The dimensions column shows the number of parameters that need to be estimated for the model.

that is set to 1. Still other programs normalise so that the sum of all six relative rates is 6 or 1, or other numbers (for example, so that the expected output of the instantaneous rate matrix is 1 – which also requires knowledge of the base frequencies). Luckily, you can easily compare two sets of relative rates that have different normalisations by (for example) rescaling one set of relative rates so that their sum (all six) is equal to the sum of (all six of) the other.

Use GTR to define other models: The GTR model can be used to define less parameter-rich models such as TN93 (Tamura and Nei 1993) or even JC69 (Jukes and Cantor 1969) by linking some of the parameters. In fact, GTR can be used to define any reversible nucleotide substitution model by linking parameters and changing the frequency estimation method. Table 7.3 shows which rates should be shared to get a particular substitution model. Note that since the rates are normalised in a BEAST analysis so that on average one substitution is expected in one unit of time, there is one parameter that can be fixed to a constant value, which is required to ensure convergence. In general, it is a good idea to fix a rate for which there is sufficient data to get a good estimate from the alignment. Most nucleotide alignments contain C–T transitions, so the default in BEAST is to fix that rate to 1 and estimate other rates relative to that.

Number of gamma categories: The number of gamma categories determines how well rate heterogeneity following a gamma distribution is approximated by its discretisation into a number of distinct categories. The computational time required by BEAST is linear with the number of gamma categories, so a model with 12 rate categories takes three times longer than one with four rate categories. For most applications four to eight categories does a sufficiently good job at approximating a gamma distribution in this context (Yang 1994). A good strategy is to start with four categories. If the estimate of the shape parameter turns out to be close to zero this is an indication that there is substantial rate heterogeneity and a larger number of categories can result in a better fit. One situation where rate heterogeneity among sites occurs is when the sequences are protein coding, and sites at codon positions 1 and 2 are known to have a lower rate than those at codon position 3 due to purifying selection on the underlying amino acid sequences and redundancy of the genetic code leading to reduced selection at codon position 3. In these cases, partitioning the data by codon position may fit better than gamma categories (Bofkin and Goldman 2007; Shapiro et al. 2006) and this model is up to x times faster than using gamma categories, where x is the number of categories.

Prior for the gamma shape parameter α: The default prior for the gamma shape parameter in BEAST is an exponential with mean 1. Suppose you would use a uniform distribution between 0 and 1000 instead. That means that the prior probability of the shape parameter α being >1 (and smaller than 1000) is 0.999 (99.9%) and the prior probability of $10 < \alpha < 1000$ is 0.99. As you can appreciate this is a fairly strong prior for almost-equal rates across sites. Even if the data have reasonably strong support for a shape of, say, 1.0, this may still be overcome by the prior distribution. All priors in BEAST are merely defaults, and it's impossible ahead of time to pick sensible defaults for all possible analyses. So it is important in a Bayesian analysis to assess your priors.

It is better to choose a good prior, then to fix the parameter value to the maximum likelihood estimate. Doing the latter artificially reduces the variance of the posterior distribution by using the data twice if the estimate came from the same data.

Invariant sites: Without invariant sites (that is, proportion invariant = 0), the gamma model acts on all sites. When setting the proportion of invariant sites to a non-zero value, the gamma model for rate heterogeneity across sites will need to explain less rate variation in the remaining variable sites. As a consequence, the remaining sites look less heterogeneous and the estimated value of the shape parameter will increase. Estimating a proportion of invariant sites introduces an extra random variable, and thus injects extra uncertainty in the analysis, which may not always be desirable (see model selection strategy in Chapter 10). The paradox is that for $\alpha < 1$ and short total tree lengths you expect sites that are invariant even though they have a non-zero rate.

Frequency model: For the HKY (Hasegawa et al. 1985), GTR and many other substitution models it is possible to choose the way frequencies are determined when calculating the tree likelihood. The three standard methods are uniform frequencies (one-quarter each for nucleotide models), empirically estimated from the alignment and estimated by MCMC. Uniform frequencies are generally only useful when you have done some simulations and uniform is the truth. Empirical frequencies are regarded as a reasonable choice, although obviously they are based on a simple average, so that if you have a lot of identical sequences at the end of a long branch then the empirical frequencies will be dominated by that one sequence which might not be representative. Estimating frequencies is the most principled approach and the estimates typically converge fast. However, estimating the frequencies does introduce a bit of uncertainty in the analysis, which can be detrimental to analysis of parameter-rich models on low-diversity data.

Fix mean substitution rate: When there are two or more partitions that share a clock model, for example, when splitting a coding sequence into its first, second and third codon positions, it can be useful to estimate relative substitution rates. This is done in such a way that the mean substitution rate is 1, where the mean is weighted by the number of sites in each of the partitions. So, for two partitions where the first is twice as long as the second, the relative substitution rate of the first partition is weighted by two-thirds, and the second partition by one-third. In this case the first mutation rate can be 0.5 and the second two to give a mean substitution rate of $0.5 \times 2/3 + 2 \times 1/3 = 1$. Note that the unweighted mean does not equal 1 in this case.

For a mean substitution rate to be fixed to 1, substitution rates of at least two partitions must be marked as estimated. Warning: when the mean substitution rate is not fixed, but substitution rates are estimated and a clock rate for the partition is also estimated, the chain will not converge since the clock and substitution rates are not identifiable. This happens because there are many combinations of values of substitution rate and clock rate that result in the same tree likelihood.

Typically, BEAST uses substitution rates for relative rates, and the clock rate for absolute rates. But in the end, the actual rate used to calculate the likelihood is the product of these two. BEAUti just makes it convenient to add a prior on substitution rates (by checking the Fix Mean Substitution Rate check box) such that the mean rate of the substitution rates across multiple partitions is one.

7.2.2 Clocks and rates

Which clock model to choose: In most cases, it is a good idea to investigate how clock-like your data are. One approach is simply to perform both a strict clock and relaxed clock analysis and let the data decide which one is most appropriate (see Section 10.1 on interpreting trace logs). If the data turn out to be non-clock-like, the other relaxed clock models can be run and model comparison (see Section 9.5) used to find the one that performs best. However, for intra-species data sets it's not a good idea to use a relaxed clock (except for some rapidly evolving viruses) for the following reasons:

Firstly, there is little reason to expect that rates of evolution are changing from lineage to lineage within a species as the life history characteristics and DNA repair mechanisms are essentially the same for all individuals.

Secondly, intra-species data sets are generally very data-poor (few haplotypes) so the fewer parameters in your model the better.

Thirdly, in intra-species data sets, the apparent rate variation visible in trees obtained from an unconstrained analysis can often be explained simply by the expected stochastic noise in the evolutionary process. For a sequence length of L, along a branch with a true length of b, you expect a Poisson random number of mutations (x) with a mean of $E[x] = b \times L$ (see Chapter 4). But its important to realise that this x will vary from one realisation of the evolutionary process to the next (one sister branch to the next), so two branches can have different maximum likelihood lengths, even if their true underlying time and mutation rates are identical. This is especially true when $E[x]$ is close to zero (see Chapter 4).

Finally, for mtDNA-like data sets, it has been shown by simulation that if the true standard deviation of log rate on branches, S, is low (< 0.1), then strict clock analyses perform well at recovering the clock rate and divergence times (Brown and Yang 2011). Note that this rule applies for the *true* value of S, which is not accessible in small data sets, where the uncertainty in the estimated S can be very large. So this rule could be applied if there is no evidence to *reject* $S < 0.1$.

In short, for low-diversity intra-species data sets there are often low levels of rate variation between branches, and in that case the strict molecular clock can be superior for rate and divergence-time estimation (Brown and Yang 2011). Even with significant evidence for moderate rate variation, it is worth using a strict molecular clock if you determine that the model misspecification is not severe (i.e. if there is no evidence to *reject* $S < 0.1$).

Absolute and relative rates: The substitution rate for a partition is the product of the substitution rate specified for the site model and the clock rate of the clock model. Care must be taken to determine which rate parameters should be estimated and which should be fixed to constant values. Failure to choose the correct combination will result in invalid analysis or nonsensical results.

If there is one partition and no timing information in the form of calibration or tip dates, both substitution rate and clock rate should be fixed to a constant value or have a strongly informative prior. Usually substitution rate is set to 1.0 and clock rate is set to either 1.0, or a known molecular clock rate. If a proper informative prior has been constructed based on the literature or previous analyses, then the clock rate can be 'estimated', in the sense that uncertainty can be incorporated into its prior. Since clock rate is a scale parameter, its prior should typically be a log-normal, rather than a normal, so that the prior probability is zero for a rate of zero.

If for a single partition there is some timing information, one of the clock rate and substitution rate should be estimated, but not both. If both substitution and molecular clock rates are estimated, the analysis will be invalid due to non-identifiability. Typically, you want the clock rate to be estimated, which is set by BEAUti by default.

Conversely, if both rates are fixed and the rate $\hat{k}$ implied by the timing information, t (in conjunction with the genetic distances d from the sequence data; i.e. $\hat{k} = d/t$) differs considerably from the specified overall clock rate then divergence-time estimation may be compromised. A mismatch of timing information and fixed molecular clock rates could also cause poor convergence.

For multiple partitions sharing the same tree, one partition can be chosen as reference and the remainder can have their substitution rates estimated relative to the reference partition. Preferably, the substitution rates of all partitions are estimated, but with the constraint that the mean substitution rate per site is fixed to 1 (Drummond 2002; Pybus et al. 2003). To calculate the mean substitution rate per site, the length of partitions is taken in account, as longer partitions will have a bigger impact on the mean substitution rate than shorter ones.

Multiple partitions do not need to share a single time-tree. For each individual tree in the analysis, molecular clock rates can be estimated if there is timing information to inform the rate. However, per tree at least one substitution rate should be fixed or alternatively the mean of the substitution rates must be fixed.

Set the clock rate: As mentioned above, when a clock rate is available from another source, like a generalised insect clock rate of 2.3% per million years, you may want to use this fixed rate instead of estimating it. A rate of 2.3% per million years equals 0.023 substitutions per site per million years (or half that if it was a divergence rate, that is the rate of divergence between two extant taxa), so when you set the clock rate to 0.023 it means your timescale will be in millions of years.

Mutation/clock rate prior: If your prior on the rate parameters were the default ones (currently, BEAUti assumes a uniform distribution with an extremely large upper bound), and your timing information is not strongly informative, then you should expect unreasonable values for rate estimates. This is because you are sampling an inappropriate default prior, which due to the lack of signal in the data will dominate the rate estimate. The uniform prior in this context puts a large fraction of the prior mass on large rates (larger than one substitution per unit of time), resulting in nonsensical estimates.

Of course if you want a normal prior in log-space then you can choose the log-normal prior. The log-normal is recommended rather than normal since the normal distribution has a non-zero density for a rate of zero, which makes no sense for a scale parameter like clock and substitution rate. A log normal with appropriate mean and 95% HPD bounds is a good choice for a substitution rate prior in our opinion. If you have a number of independent rate estimates from other papers on relevant taxa and gene regions, you can simply fit a log-normal distribution to them and use that as a prior for your new analysis.

Alternatively, a diffuse gamma distribution (shape 0.001, scale 1000) can be appropriate. In general, if the data are sufficiently informative and there is a good calibration, the prior should not have too much impact on the analysis. The initial value should not matter, but setting it as the mean of the distribution does no harm.

There is a reasonably good (but broad) prior on the rate of evolution available for almost any group of organisms with a careful read of the literature. For example, vertebrate mitochondrial evolution tends to be in the range 0.001–1 substitutions per site per million years. While this is a broad prior it certainly rules out more than it rules in. However, if you want to assert ignorance about the clock rate parameter it is fine to take a broader prior on the clock rate as long as there is a proper prior on one of the divergences in the tree and a proper prior on the standard deviation of the log rate S if you use a log-normal relaxed clock. The default prior for S (exponential with mean $\frac{1}{3}$) is suitable for this purpose, although a smaller mean (e.g. 0.1) of the exponential prior on S for closely related taxa can be easily argued for.

7.2.3   Tree priors

Which tree prior to choose: It always makes sense to investigate the sensitivity of your estimates to your choice of prior, so we recommend trying more than one tree prior if you are uncertain which one to use. If you have a single sequence from each species then the Yule process is the simplest tree prior to choose in the first instance. The Yule model is a simple pure birth process with a single parameter, the birth rate λ. Including the potential for both lineage birth and lineage extinction gives the two-parameter constant-rate birth–death model, which has been suggested to be an appropriate 'null model' for phylogenetic diversification (Mooers and Heard 1997; Nee 2001; Nee et al. 1994). However, unless all extant species are included in the phylogeny, an additional sampling fraction, representing the proportion of extant species present in the phylogeny, should

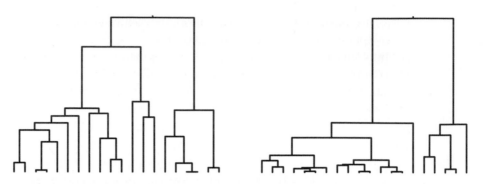

Figure 7.1 Left, a simulated Yule tree; right, a simulated coalescent (with constant population) tree with 20 taxa. Note, coalescent trees have much shorter branches near the tips.

also be included as a random variable in the model (Stadler 2009; Yang and Rannala 1997). See Chapter 2 for details on these tree priors.

The sampling proportion can make a big difference to the prior if it is very small (Stadler 2009). But if we put the sampling proportion aside for a moment then Yule versus birth–death is not as big a difference in priors as Yule versus coalescent. The coalescent prior varies quadratically with the number of lineages spanning an inter-node time $(E[\Delta t_k] \propto \exp\left(-\binom{k}{2}\right)$ where Δt_k is the length of time spanned by the kth interval in the tree) whereas the Yule prior varies linearly $(E[\Delta t_k] \propto \exp(-k))$ (see Figure 7.1).

The Yule prior assumes that the birth rate of new lineages is the same everywhere in the tree. For example, it would assume that the birth rate of new lineages within the set of taxa of interest is the same as the rate of lineage birth in the outgroup. This assumption may not be appropriate. A relaxed molecular clock also allows quite some scope for changes in the rate from branch to branch, so that the molecular clock model can interact with the tree prior to accommodate a poor fit of the tree prior by compensating with lineage-specific rate variation.

The coalescent is a tree prior for time-trees relating a small sample of individuals from a large background population, where the background population may have experienced changes in population size over the time period that the tree spans. Non-parametric forms of the coalescent tree prior are more flexible and so may accommodate a wider range of divergence time distributions. However, the parameters of the coalescent tree prior will be difficult to interpret if your sequences come from different species.

The skyline (Drummond et al. 2005) and skyride (Minin et al. 2008) priors are coalescent priors that are useful for complex dynamics. The parameterised coalescent priors like the constant and exponential coalescent are useful when you want to estimate specific parameters, for example the growth rate. However, you have to be confident that the model is a good description of the data for such estimates to be valid.

Note that some priors, such as the Yule prior, assume all tips are sampled at the same time. These priors are not available for use with serially sampled data, and the birth–death skyline plot can be used instead (Stadler et al. 2013).

For divergence-time dating, once a tree prior has been chosen, we highly recommend to practitioners that they look carefully at the resulting prior distribution (running the analysis without data) on node ages to see whether the prior reflects prior beliefs about possible times of origin of clades of interest (see also Section 7.2.4 on calibrations).

Prior for Yule birth rate: If you are using the Yule tree prior then the birth rate λ needs to be specified. This parameter governs the rate at which species diverge. This rate, in turn, determines the (prior) expected age of the species tree (denoted by t_{root}). The formula connecting these is two quantities is

$$\lambda = \frac{1}{t_{root}} \left(\sum_{k=1}^{n-1} \frac{k}{n(n-k)} \right), \tag{7.2}$$

where n is the number of species. So, if you have some information about the speciation rate, you can use this to specify a prior on λ. This will effectively form a prior distribution on the age of the tree through the tree prior.

If no prior information on the speciation rate λ is available, then a uniform prior is okay. Note that in practice for *BEAST analysis only we found that $1/x$ works better for a birth rate prior on the species tree. Such a prior puts higher preference on lower birth rates, hence older species trees. But the age of the species tree is limited from above by corresponding divergences in the gene trees, which each have their own coalescent prior. As a result the species tree will 'snugly' fit the gene trees, while with a uniform birth rate prior it might linger lower than the gene trees would justify.

Prior for constant population size: When using the constant-size coalescent tree prior, a choice needs to be made for the prior on the constant population size hyperparameter. If you have prior information, then an *informative* prior is always more preferable.

Otherwise, the $1/x$ prior (which for the population size parameter is also the Jeffrey's prior) is a good non-informative prior for population size. We showed in Table 3 of (Drummond et al. 2002) that this prior leads to good recovery of the population size in terms of frequentist coverage statistics, but we were not attempting to provide a general result in that paper. It is well known that estimation of the growth rate for exponentially growing populations is positively biased (especially when using a single locus, see Kuhner et al. 1998) and we do not know of any work that has been done on appropriate priors for that parameter. Having said $1/x$ is okay for θ, it is not if you want to do path sampling (see Section 9.5), because path sampling is only possible when the prior is proper, whereas the $1/x$ prior is improper since it doesn't have a finite integral.

Number of BSP groups: The number of groups for a BSP prior determines how well the demographic function can be approximated. When there is only a single group, BSP becomes equivalent to a simple constant population size model. The maximum number of groups is one per coalescent event and in this case a population size is estimated for every interval between two coalescent events. Choosing the maximum number of groups leads to a large number of estimated parameters and these estimates will be very noisy. So, the choice for the number of groups typically needs to be in between these two extremes. Too few groups results in loss of signal of the population size history, and too many groups result in noisy estimates that are difficult to interpret.

The optimal value depends on the data at hand and the demographic history being inferred. A rule-of-thumb is to start with five groups and if more resolution is required increase the number of groups in subsequent runs. Alternatively, use EBSP, which supports estimation of the number of change-points from the data as well as use of multiple loci (Heled and Drummond 2008; see Section 2.3.1);

Different behaviour for different tree hyper-priors: Different tree hyper-parameter priors can result in differences in rate estimates. For example, a uniform prior on Yule birth rate works the way one expects, while the uniform prior on the population size of a constant-size coalescent prior does not. The reason for this lies in the way in which the different models are parameterised. If the coalescent prior had been parameterised with a parameter that was equal to 1 over the constant population size, then a uniform prior would have behaved as expected (in effect the Jeffrey's prior is performing this re-parameterisation). Conversely, if the Yule tree model had been parameterised by the mean branch length (for derivation of mean branch length in a Yule tree see Steel and Mooers 2010) it would have behaved in a similar way to coalescent prior with a uniform prior on the population size.

Before you start thinking that we parameterised the coalescent prior incorrectly, it is important to realise there is no parameterisation that is correct for all questions. For some hypotheses one prior distribution is correct, for others another prior distribution works better. The important thing is to understand how the individual marginal priors interact with each other. For an analysis involving divergence-time dating and rate estimation one should be aware that the tree prior has the potential to influence the rate estimates and vice versa.

Finally, if you have strongly informative prior distributions on divergence times or rates (like gamma or log-normal distributions with moderate standard deviations) then most of these effects become negligble. But refer to Heled and Drummond (2012, 2013) for a detailed discussion of calibrated tree priors.

7.2.4   Calibrations

Choosing a prior for a calibration: If you have some idea about where most of the probability for a calibration may be, for example, you are 95% sure that the date must be between X and Y, or you are 95% sure that the divergence date is not older than X, then BEAUti can help in choosing the correct prior. BEAUti can display a probability density for a variety of potential prior distributions, and also report the 2.5% and 5% upper and lower tails of the distribution. So, you can select the parameters of a prior by matching the 90% or 95% bounds of your prior information with the tails of the chosen prior distribution.

Note that a calibration density based on the estimated divergence time of a previous analysis of the same data will lead to inappropriately small credible intervals for the estimated divergence ages. Using the data twice (double-dipping) leads to credible intervals that no longer accurately represent the true uncertainty under the model. But it should be noted that using data to define priors is in certain circumstances regarded as reasonable, especially when employed in *empirical Bayes approaches* which have been used successfully in some phylogenetic contexts (Huelsenbeck et al. 2001; Nielsen and Yang 1998; Yang et al. 2000).

An alternative is to base priors on an independent data set. This can be effective when there is not enough temporal information in the data set you are working with. The estimate of the clock rate of the first data set can be used to put a prior on the molecular clock rate of the data set of interest. You can then approximate this distribution with a normal or gamma distribution and use it as an independent empirically derived prior. However, it is important to make sure the data sets are independent and have no sequences in common and the samples are indeed independent, that is, the sequences from the data sets are not para- or polyphyletic in a combined phylogeny.

Divergence times are scale parameters, so they are always a positive number. Therefore you should use priors that are appropriate for scale parameters like log-normal which is a density that is only defined for positive values (see Table 7.2 for other distributions). As a rule, normal distributions should not be used for calibration densities. Divergence times are defined only on the positive number line, and so your prior should be as well.

Another method for determining the shape of the calibration density is through the CladeAge package (Matschiner and Bouckaert 2013). It requires specification of the net diversification rate, turnover rate and fossil sampling rate. Using this information, together with the interval that specifies the geological age range of the fossil, a simulation-based calibration density can be generated.

The calibrations, monophyly constraints and the tree prior can interact with each other in unpredictable ways. For example, a calibration on the MRCA time of a clade limiting its height has an impact on any calibration on a subclade that has tails exceeding the limit of the parent clade (Heled and Drummond 2012, 2013). Therefore, you should use a calibrated tree prior wherever possible when doing divergence time dating in BEAST (Heled and Drummond 2012, 2013). If that is not feasible then always run your analysis without data to see how the tree prior and calibration densities interact. This can be done by selecting the sample from prior option in BEAUti. This way, you can determine what the actual marginal prior distributions are for divergence times of interest. Calibration densities combined naively with the tree prior (Yule, or coalescent) can yield surprising results, and you might need to change your calibrations accordingly.

When calibrating an analysis with a single calibration density and using the Yule prior, the 'calibrated Yule prior' can be used (Heled and Drummond 2012) that will guarantee that the marginal distribution of the calibrated node corresponds to the calibration density. With more than one calibration, the calibrated birth–death prior implemented in BEAST can become computationally burdensome, and this is an active area of research (Heled and Drummond 2013). A superior alternative for fossil dating has very recently been developed that can handle large numbers of fossils under the birth–death sampling model, including estimation of the correct phylogenetic location for each fossil taxa (Gavryushkina et al. 2014).

Multiple calibrations: When using the strict clock, a single calibration tends to be sufficient, assuming there are sufficient sequence data. If you decide to use multiple calibrations, then BEAST, as always, samples the tree topology and branch lengths that give good posterior probabilities, that is, it samples a region of the parameter space proportional to its posterior probability. If multiple calibrations do not fit well with each other in light of the sequence data, then some kind of compromised inference will result. This could mean that some of the marginal posterior distributions of calibrated divergence times end up far away from their prior distributions. As noted before, unless carefully designed, multiple calibrations can interact with each other, so make sure that the joint effect of the calibrations represent your prior information by running BEAST without data. Alternatively, consider using BEAST's implementation of the fossilised birth–death prior for calibration (Gavryushkina et al. 2014; Heath et al. 2014).

Calibration and monophyly: Depending on the source of information used for informing a calibration, often the clade to which the calibration applies is assumed to be monophyletic (the oldest fossil penguin can't be assigned to an ancestral divergence in the tree, it is not precisely clear which taxa constitute the penguins). Furthermore, if the calibrated clade is not constrained to be monophyletic, the MCMC chain may mix poorly. When a taxon moves into the calibrated clade during the MCMC chain, the tree prior and posterior may prefer an older age for the clade, whereas when the taxon moves out again, the opposite will be true. Moving between these different modes can pose a problem for the standard MCMC operators.[1] For both of these reasons we recommend constraining calibrated nodes to be monophyletic when it can be justified.

Calibrating the age of the root vs. fixing the molecular clock rate: It can happen that there is no timing information, or the timing information is of no interest to the analysis, for example, if only the tree topology, or relative divergence times are required, or the geographical origin in a phylogeographical analysis is of interest. It is tempting to put a calibration on the root of the tree, perhaps with a small variance. However, if timing is of no interest, it might be better to fix the clock rate to 1.0 because the MCMC chain will mix more efficiently that way. This is because some operators act to change the topology by changing the root node age, which is hampered by a tight calibration on the root. Fixing the clock rate to 1.0 will not interfere with the efficiency of the tree operators and will result in branch lengths that have units of expected substitutions.

7.3   Miscellanea

Installing and managing packages: Packages, also known as plug-ins or add-ons, contain extensions of BEAST, such as the reversible jump substitution model. For desktop computers, installing and managing packages is best done through BEAUti, which has a package management dialogue under the File → Manage packages menu. There is a command line utility called `packagemanager` to manage packages on server

---

[1] But if needed mixing problems like this can be addressed using Metropolis-coupled MCMC.

computers. More information for specific operating systems and on where packages are installed is available on the BEAST wiki.

Log frequency: In order to prevent log files from becoming too large and hard to handle, the log frequency should be chosen so that the number of states sampled to the log file does not exceed 10 000. So, for a run of 10 million, choose 1000 and for a run of 50 million, choose 5000 for the log frequency. Note that to be able to resume it is a good idea to keep log intervals for trace logs, tree logs and the number of samples between storing the state all the same. If they differ and an MCMC run is interrupted, log files can be of different lengths and some editing may be required (see below).

Operator weights: There can be orders of magnitude difference between operator weights. For example, the default weight for the uniform operator which changes divergence times in a tree without changing the topology is 30, while the default weight for the exchange operator on frequencies is just 0.01. This is because in general the age of divergences in trees is hard to estimate hence requires many moves, while the frequencies for substitution models tend to be strongly driven by alignment data in the analysis. In BEAUti, you can change operator weights from the defaults. Typically, the defaults give reasonable convergence for a wide range of problems. However, if you find that ESSs are low for some parameters while many others are high, generally you want to increase the weight on operators acting on parameters with low ESSs in order to equalise out the mixing of different components of the posterior. Weights for operators on the tree should increase if the number of taxa in the tree increases. Whether the increase should be linear or sub-linear requires further research.

Fixed topology: The topology can be kept constant while estimating other parameters, for example divergence times. This can be done in BEAUti by setting the weights of operators that change the topology of the tree to zero. Alternatively, the operators can be removed from the XML. The standard operators that change the tree topology are subtree-slide, Wilson–Balding and the narrow and wide exchange operators (for information about tree proposals, see Höhna and Drummond 2012). For *BEAST analysis, the node-reheight operator affects the topology of the species tree (Heled and Drummond 2010).

Newick starting tree: A user-defined starting tree can be provided by editing the XML. Using the `beast.util.TreeParser` class, a tree in Newick format can be specified. There are a few XML files in the examples directory that show how to do this. BEAST only accepts binary trees, so if your tree has polytomies you have to convert the tree to a binary tree and create an extra branch of zero length. For example, if your polytomy has three taxa and one internal node, say (A:0.3,B:0.3,C:0.3), then the binary tree ((A:0.3,B:0.3):0.0,C:0.3) is a suitable representation of your tree.

Multi-epoch models: When there is reason to believe that the method of evolution changes at different time intervals, for example because part of the time frame is governed by an ice age, a single substitution model may not be appropriate. In such a situation, a multi-epoch model (Bielejec et al. 2014) can be useful. In the `BEASTlabs` package there is an implementation of a `EpochSubstitutionModel` where you can specify different substitution models at different time intervals.

7.4   Running BEAST

Random number seed: When starting BEAST, a random number seed can be specified. Random number generators form a large specialised topic (Knuth 1997). A good random number generator should not allow you to predict what a sequence will be for seed B if you know the sequence for seed A (even if A and B are close). BEAST uses the Mersenne Prime twister pseudo-random number generator (Matsumoto and Nishimura 1998), which is considered to be better than the linear congruential generators that are the default in many programming languages, including Java. A pseudo-random number generator produces random numbers in a deterministic way. That means that if you run a particular version of BEAST with the same seed and the same input file, the outcome will be exactly the same. It is a good idea to run BEAST multiple times so that it can be checked these runs all converge on the same estimates. Needless to say, the runs should be started with different seeds, otherwise the runs will all show exactly the same result.

Which seed you use does not matter and seeds that only differ by one result in completely different sequences of random numbers being generated. By default, BEAST initialises a seed with the number of milliseconds since 1970 according to the clock on your computer at the time you started BEAST. If you want a run you can exactly reproduce you should override this with a seed number of your choice.

Stopping early: Once a BEAST run is started, the process can be stopped due to computer failure or by killing the process manually. If this happens, you should check that the log files were properly completed, because the program might have stopped in the middle of writing a line to the file. If a log line is corrupted, the partial line should be removed. For some post-processing programs like Tracer, when doing a demographic reconstruction it is important for the tree file to be properly finalised with a line containing 'End;'. Failing to do so may result in the analysis being halted.

Resuming runs: A state file representing the location in sample space and operator tuning parameters can be stored at regular intervals so that the chain can be resumed when the program is interrupted due to a power failure or unexpected computer outage. BEAST can resume a run using this state file, which is named the same as the BEAST XML input file name with `.state` appended to the end (e.g. `beast.xml.state` for `beast.xml`). When resuming, the log files will be extended from the point of the last log line. Log files need to end in the same sample number, so it is recommended that the log frequency for all log and tree files is kept the same, otherwise a run that is interrupted requires editing of the log files to remove the last lines so that all log files end with the same sample number as the shortest log. The same procedure applies when a run is interrupted during the writing of a log file and the log files became corrupted during that process.

BEAGLE: BEAGLE (Ayres et al. 2012; Suchard and Rambaut 2009) is a library for speeding up phylogenetic likelihood calculations that can result in dramatic improvements in run time. You have to install BEAGLE separately, and it depends on your hardware and data how much performance difference you will get. BEAGLE can utilise some types of graphics processing units, which can significantly speed up phylogenetic likelihood calculations, especially with large data sets or large state spaces like codon models. There are a considerable number of BEAGLE options that can give a performance boost, in particular switching off scaling when there are not a large number of taxa, choosing single precision over double precision calculations and choosing the SSE version over the CPU version of BEAGLE. These options may need some extra command line arguments to BEAST. To see which options are available, run BEAST with the -help option. It requires a bit of experimentation with different BEAGLE settings to find out what gives the best performance. It actually depends on the data in the analysis and the hardware you use.

Note that some models, such as the multi-epoch substitution model and stochastic Dollo model (Nicholls and Gray 2008) and multi-state stochastic Dollo (Alekseyenko et al. 2008), are not supported by BEAGLE. Also, the tree-likelihood for SNAPP (Bryant et al. 2012) is not currently supported by BEAGLE.

8   Estimating species trees from multilocus data

The increasing availability of sequence data from multiple loci raises the question of how to determine the species tree from such data. It is well established that just concatenating nucleotide sequences results in misleading estimates (Degnan and Rosenberg 2006; Heled and Drummond 2010; Kubatko and Degnan 2007). There are a number of more sophisticated methods to infer a species phylogeny from sequences obtained from multiple genes. This chapter starts with an example of a single locus analysis to highlight some of the issues, then details the multispecies coalescent. The remainder describes two multilocus methods for inferring a species phylogeny from DNA and SNP data respectively. Though even multispecies coalescent may suffer from detectable model misspecification (Reid et al. 2013), it has not been shown that it is worse than concatenation.

8.1   Darwin's finches

Consider the situation where you have data from a single locus, but have a number of gene sequences sampled from each species and you are interested in estimating the species phylogeny. Arguably, even in this case, an approach that explicitly models incomplete lineage sorting is warranted. The ancestral relationships in the species tree can differ considerably from those of an individual gene tree, due (among other things) to incomplete lineage sorting. This arises from the fact that in the absence of gene flow the divergence times of a pair of genes sampled from related species must diverge earlier than the corresponding speciation time (Pamilo and Nei 1988). More generally, a species is defined by the collection of all its genes (each with their own history of ancestry) and analysing just a single gene to determine a species phylogeny may therefore be misleading, unless the potential discrepancy between the gene tree and the species tree is explicitly modelled.

For example, consider a small multiple sequence alignment of the mitochondrial control region, sampled from 16 specimens representing four species of Darwin's finches. The variable columns of the sequence alignment are presented in Figure 8.1.

The alignment is composed of three partial sequences from each of *Camarhynchus parvulus* and *Certhidea olivacea*, four from *Geospiza fortis* and six from *G. magnirostris* (Sato et al. 1999). The full alignment has 1121 columns and can be found

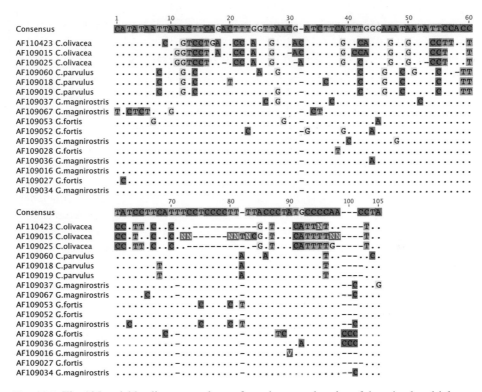

Figure 8.1 The 105 variable alignment columns from the control region of the mitochondrial genome sampled from a total of 16 specimens representing four species of Darwin's finches. The full alignment is 1121 nucleotides long.

in the examples/nexus directory of the BEAST2 distribution. To analyse these data we use an implementation of the multispecies coalescent (known as *BEAST; see Section 8.3 for details) with the HKY substitution model and a strict molecular clock. Figure 8.2 shows a summary of the posterior distributions for the gene tree and the species tree.

The most probable species tree, shown at the bottom of Figure 8.2, is supported by about two-thirds of the posterior. The alternative topologies are shown with intensity proportional to their support at the top of the figure. This is despite the fact that in the gene tree a clade containing all the *Geospiza* sequences has almost 100% support. There is some uncertainty to whether *C. parvulus* descended from the root but almost none on the monophyly of the *Geospiza* sequences. This example is especially illustrative of how the gene tree is not necessarily the same as the species tree, even if there is only a single gene in the analysis. The taxon labels are coloured by species in the gene tree in Figure 8.2. So, though the sequences for *G. fortis* and *G. magnirostris* together form a monophyletic clade in the summary gene tree, the species do not fall into two clear monophyletic subclades. In fact, the *G. magnirostris* and *G. fortis* sequences are mixed together in an arrangement known as *paraphyly*, and there is no obvious species boundary visible from direct inspection of the gene tree.

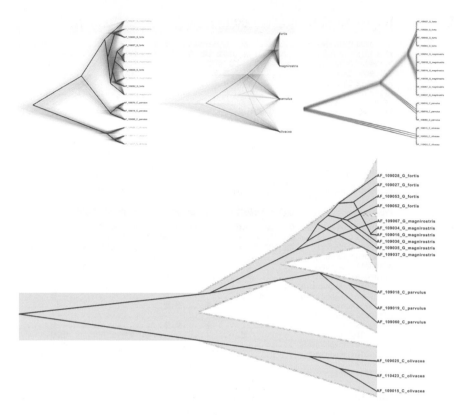

Figure 8.2 Single locus analysis of Darwin's finches. Top left, gene trees, top middle species trees, top right shows the gene trees when the species tree is fixed and sampling from the prior, which shows the prior puts a preference on trees that are as low as the species tree allows. Bottom shows how the summary tree of the gene trees fits in the species tree. Branch width of the species tree indicates population sizes.

A standard non-*BEAST analysis of these data results in approximately the same tree as the gene tree shown in Figure 8.2.

When the species tree and population sizes are fixed, a sample from the prior distribution on gene trees is possible in a *BEAST analysis. Figure 8.2 shows three gene tree samples from this prior, with the species tree set to the median estimate from a posterior analysis of the Darwin's finches data. It can be seen that with the estimated divergence times and population sizes, paraphyly is expected for the two *Geospiza* species, whereas monophyly is the tendency for both *C. olivacea* and *C. parvulus*.

The posterior point estimate of the gene tree enclosed in the corresponding species tree estimate is shown at the bottom of Figure 8.2. This shows the summary tree for the gene tree fitted in the summary tree for the species tree. Branch widths of the species tree indicate population sizes estimated by *BEAST, showing increase in population sizes with time going forward for all the species. The clades for *C. parvulus* and *C. olivacea* are each monophyletic, lending support for the older estimates of their corresponding speciation times relative to population sizes.

8.2   Bayesian multispecies coalescent model from sequence data

In an empirical study (Leaché and Rannala 2011) it was shown Bayesian methods for species tree estimation perform better than maximum likelihood and parsimony. In a Bayesian setting the probability of the species tree S given the sequence data (D) can be written (Heled and Drummond 2010):

$$f(\mathbf{g}, S|D) = \frac{f(S)}{\Pr(D)} \prod_{i=1}^{m} \Pr(D_i|g_i)f(g_i|S), \tag{8.1}$$

where $D = \{D_1, \ldots, D_m\}$ is the set of m sequence alignments, one for each of the gene trees, $\mathbf{g} = \{g_1, \ldots, g_m\}$. The term $\Pr(D_i|g_i)$ is a standard tree likelihood, which typically subsumes a substitution model, site model and clock model for each of the individual genes (see Chapters 3 and 4 for details). $f(g_i|S)$ is the multispecies coalescent likelihood, which is a prior on a gene tree given the species tree and $f(S)$ is the prior on the species tree (see Section 2.5.1 for details).

The species tree prior $f(S)$ can be thought of as consisting of two parts: a prior on the species time-tree (g_S), $f(g_S)$, and a prior on population sizes, $f(\mathbf{N})$, together giving $f(S) = f(g_S)f(\mathbf{N})$. The species time-tree prior $f(g_S)$ is typically the Yule or birth–death prior (see Chapter 2 for details).

In order to estimate a species tree with the posterior distribution in Equation (8.1) one approach is to simply sample the full state space of gene trees and species tree using MCMC and treat the gene trees ($\mathbf{g}$) as nuisance parameters, thereby summarising the marginal posterior distribution of the species tree. This effectively integrates the gene trees out via MCMC (Heled and Drummond 2010; Liu 2008):

$$f(S|D) = \frac{f(S)}{\Pr(D)} \int_G \prod_{i=1}^{m} \Pr(D_i|g_i)f(g_i|S)dG. \tag{8.2}$$

An alternative approach that has been used in the application of the multispecies coalescent to SNP data (Bryant et al. 2012) is to numerically integrate out the gene trees for each SNP so that:

$$f(S|D) = \frac{f(S)}{\Pr(D)} \prod_{i=1}^{m} \Pr(D_i|S), \tag{8.3}$$

where $\Pr(D_i|S) = \int_{G_i} \Pr(D_i|g_i)f(g_i|S)dG_i$ (Bryant et al. 2012).

8.3   *BEAST

BEAST 2 includes a Bayesian framework for species tree estimation. The statistical methodology described in this section is known by the name *BEAST (pronounced 'star-beast', which is an acronym for Species Tree Ancestral Reconstruction using BEAST) (Heled and Drummond 2010). The model assumes no recombination within

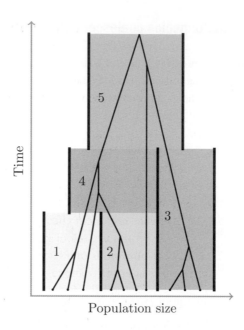

Figure 8.3 A species tree with constant size per species branch. For n_S species this leads to $2n_S - 1$ population size parameters.

each locus and free recombination between loci. Approaches that include hybridisation to the species tree are in development (Camargo et al. 2012; Chung and Ané 2011; Yu et al. 2011). A tutorial is available for *BEAST that uses three gopher genes (Belfiore et al. 2008) to estimate the species tree (see BEAST wiki).

*BEAST does not require that each gene alignment contains the same number of sequences. It also does not need the same individuals to be sampled for each gene, nor does it need to match individuals from one gene to the next. All that is needed is that each sequence in each gene alignment is mapped to the appropriate species. Note that *BEAST cannot be used with time-stamped sequences at the time of writing due primarily to technical limitations in the implementation of the MCMC proposals. For details on the *multispecies coalescent* model that underlies *BEAST, see Section 2.5.1.

Most multispecies coalescent models assume that the population size is constant over each branch in the species tree (Figure 8.3). However, two other models of population size history are implemented in *BEAST. The first allows linearly changing population sizes within each branch of the species tree including the final ancestral population at the root (see Figure 8.4). The second also allows linear changing population sizes, but has a constant population size for the ancestral population stemming from the root (see Figure 8.2 for an example of this latter option). The linear model is the most general implemented in *BEAST. The other two models can be used when fewer data are available.

The population sizes prior $f(\mathbf{N})$ depends on the model used. For constant population per branch (see Figure 8.3), the population size is assumed to be a sample from a gamma distribution with a mean 2ψ and a shape of α, that is, $\Gamma(\alpha, \psi)$ (defaults to $\alpha = 2$ at

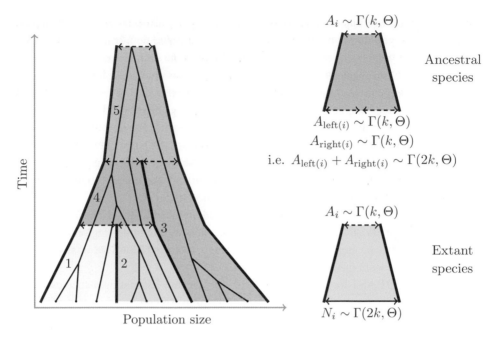

Figure 8.4 The population size priors on the branches of a three-species tree.

the time of writing). Unless we have some specific knowledge about population size, an objective Bayesian-inspired choice might be $f_\psi(x) \propto 1/x$ for hyper-parameter ψ, although note that this choice may be problematic if the marginal likelihood needs to be computed by path sampling.

In the continuous linear model, we have n_s population sizes at the tips of the species tree, and two per each of the $(n_s - 1)$ internal nodes, expressing the starting population size of each of the descendant species (Figure 8.4). The prior for the population sizes at the internal nodes are as above, but for the ones at the tips they are assumed to come from a $\Gamma(2k, \psi)$ distribution. This is chosen to assure a smooth transition at speciations because $X_1, X_2 \sim \Gamma(k, \psi)$ implies $X_1 + X_2 \sim \Gamma(2k, \psi)$. This corresponds to having the same prior on all final (most recent) population sizes of both extant and ancestral species (see Figure 8.4).

*BEAST has been applied to determine that polar bears are an old and distinct bear lineage (Hailer et al. 2012), to distinguish between single and dual origin hypothesis of rice (Molina et al. 2011), analysing the speciation process of forest geckos (Leaché and Fujita 2010) and examining cryptic diversity in butterflies (Dincă et al. 2011).

8.3.1   Performing a *BEAST analysis

The easiest way to set up a *BEAST analysis is using the StarBeast template, which you can choose from the File/Template menu in BEAUti. An extra 'Taxon Set' and 'Multi Species Coalescent' tab appears, the first for specifying which taxon from the gene tree is part of a certain species. At the multispecies coalescent tab you can choose the population function (constant, linear or linear with constant root). Also, for each of the gene trees you can specify the ploidy of the sequences.

Convergence: Note that *BEAST analysis can take a long time to converge, especially when a large number of gene trees are involved. The usual methods for speeding up tree-likelihood calculations such as using threads and fine-tuning with BEAGLE applies, as well as specifying as much prior information as possible. If few loci are available, you can run a *BEAST analysis with just a few lineages (say 20) for an initial run in order to determine model settings. Reducing sequences speeds up the chain considerably without affecting accuracy of estimates too much, since adding more sequence is not as effective as adding more loci.

Linking trees: If the genes are linked they should only be represented by a single tree in *BEAST. So genes from the same 'non-recombining' region of the Y chromosome should be represented by only a single tree in *BEAST (or any other multispecies coalescent method for that matter).

***BEAST for species tree with single gene:** Estimating a species tree from a single gene tree is perfectly valid. In fact, we highly recommend it, as it will give much more realistic assessments of the posterior clade supports. That is, it will correctly reduce the level of certainty on species tree groupings, since incomplete lineage sorting may mean the species tree is different from the gene tree, even in the face of high posterior support for groupings in the gene tree topology.

Visualising trees: DensiTree can be used to visualise a species tree where the branch widths represent population sizes. There are tools in the biopy package that can help visualise species trees and gene trees as well.[1] The *BEAST tutorial has some examples on visualising species trees.

Population size estimates: In terms of accuracy, the topology of the species tree is typically well recovered by a *BEAST analysis, the time estimates of the species tree contain larger uncertainty and population size estimates contain even larger uncertainty. The most effective way to increase accuracy of population size estimates is to add more loci (Heled and Drummond 2010). As a result, the number of lineage trees increases, and with it the number of coalescent events that are informing population size estimates. Increasing sequence lengths helps in getting more accurate time estimates, but since the number of coalescent events does not increase, population size estimates are not increasing in accuracy as much.

Species assignment: In a *BEAST analysis, it is assumed that you know the species that a sequence belongs to. However, if it is uncertain whether a sequence belongs to species A or species B, you can run two analyses, one with each assignment. The chain with the best marginal likelihood (see Section 9.5) can be assumed to contain the correct assignment (Grummer et al. 2014).

It can happen that the species assignment is incorrect, which results in coalescent events higher up in the species tree than if the species assignment is correct. As a result population size estimates will be unusually high. So, relative high population

---

[1] Available from http://code.google.com/p/biopy/.

size estimates may indicate incorrect assignments, but it can also indicate the existence of a cryptic species in your data.

Note that a species assignment of lineages does not enforce monophyly of the lineages belonging to a single species, as shown in Figure 8.2.

8.4   SNAPP

SNAPP (SNP and AFLP Package for Phylogenetic analysis) is a package in BEAST to perform MCMC analysis on SNP and AFLP data using the method described in Bryant et al. (2012). It calculates Equation (8.1) just as for a *BEAST analysis, but works with binary data such as SNP or amplified fragment length polymorphism (AFLP) instead of nucleotide data typically used in *BEAST analysis. If we call the sequence values 'green' and 'red', this means we have to specify a substitution rate u for going from green to red, and a substitution rate v for going from red to green. Another difference with *BEAST is that instead of keeping track of individual gene trees, these are integrated out using a smart mathematical technique. So, instead of calculating the integral as in Equation (8.2) by MCMC, the integral is solved numerically. This means that the individual gene trees are not available any more, as they are with *BEAST. A coalescent process is assumed with constant population size for each of the branches, so with every branch a population size is associated.

A common source of confusion, both for SNAPP and similar methods, is that the rates of mutation and times are correlated, so we typically rescale the substitution rates such that the average number of mutations per unit of time is one. Let μ be the expected number of mutations per site per generation, g the length of generation time in years, N the effective population size (number of individuals) and θ the expected number of mutations between two individuals. For a diploid population, $\theta = 4N\mu$. If μ is instead the expected number of mutations per site per year then we would have $\theta = 4N\mu g$. In the analysis in (Bryant et al. 2012) time is measured in terms of expected number of mutations. Hence

- the expected number of mutations per unit time is one;
- a branch of length τ in the species tree corresponds to τ/μ generations, or $\tau g/\mu$ years;
- the backward and forward substitution rates, u and v, are constrained so that the total expected number of mutations per unit time is one, which gives

$$\frac{v}{u+v}u + \frac{u}{u+v}v = 1;$$

- the θ values are unaffected by this rescaling. If the true mutation rate μ is known, then the θ values returned by the program can be converted into effective population sizes using $N = \theta/(4\mu)$.

There is no technical limitation to using SNAPP for divergence date estimation using either calibrations or serially sampled data. However, this has not been formally tested yet.

8.4.1 Setting up a SNAPP analysis

To set up a SNAPP analysis, first you have to install the SNAPP package if it is not already installed (see Section 7.3). Once SNAPP is installed, you need to change to the SNAPP template in BEAUti to set up a SNAPP analysis. More practical information including screenshots of BEAUti is contained in 'A rough guide to SNAPP' (Bouckaert and Bryant 2012), available via the BEAST wiki.

Mutation model settings: There is a simple way to determine the values of u and v from the data, so that they do not need to be estimated by MCMC. This has the advantage of reducing some uncertainty and calculation time and appears to have little effect on the accuracy of the analysis. Given the rates u and v, the stationary frequencies of the two states are $\pi_0 = v/(u+v)$ and $\pi_1 = u/(u+v)$, while the substitution rate is $\mu = \pi_0 u + \pi_1 v = 2uv/(u+v)$. Hence, given an estimate for π_1, and the constraint that the substitution rate is $\mu = 1$, we can solve for u and v as

$$v = \frac{1}{2\pi_1}, \quad u = \frac{1}{2\pi_0} = \frac{1}{2(1-\pi_1)}.$$

For haploid data, to get an estimate for π_1 (hence $\pi_0 = 1 - \pi_1$) count the number of ones in the alignment and divide by the number of observed sites (which equals the number of sequences times sequence length, from which the number of unknown sites are subtracted). For diploid data, count the number of ones and add twice the number of twos then divide by twice the number of observed sites.

If you prefer to estimate the substitution rates by MCMC, a hyper-prior on the rate needs to be specified. Since SNAPP assumes the substitution rate equals one, which constrains the value of u if v is known, if you estimate u this implies you estimate v as well. It can be useful to specify upper and lower bounds on the rates. In general, the MCMC for u and v should converge quickly and the estimates should have little variance.

SNAPP has a few optional extras when specifying the mutation model. You can indicate whether to use non-polymorphic data in the sequences. If true, constant sites in the data will be used as part of the likelihood calculation. If false (the default) constant sites will be removed from the sequence data and ascertainment bias is incorporated in the likelihood.

You can indicate conditioning on zero mutations, except at root (default false). As a result, all gene trees will coalesce in the root only, and never in any of the branches.

It is possible to indicate whether alleles are dominant; however, in our experience this only leads to much longer run times without changing the analysis significantly. Therefore it is assumed alleles are not dominant by default.

Tree prior settings: Just like for a *BEAST analysis, a Yule prior is appropriate for the species tree for most SNAPP analysis (see Section 7.2.3).

Rate prior settings: The SNAPP template suggests four priors for the rates: gamma, inverse gamma, CIR and uniform. Table 8.1 lists which parameters are used in which prior. The parameters for the gamma, inverse gamma and uniform distributions are the same over the entire tree, and for all trees, and the prior distribution for each ancestral population is independent. When using 'inverse gamma' as a prior we assume

Table 8.1 Parameters and their usage in the SNAPP prior

Prior	α	β	κ
Gamma	Shape	Scale	Ignored
Inverse gamma	Shape	Scale	Ignored
CIR	See text	See text	See text
Uniform	Ignored	Ignored	Ignored

independent inverse gamma distributions for thetas, so $(2/r)$ has an inverse gamma (α, β) distribution. That means that r has density proportional to $1/(r^2) * \Gamma^{inv}(2/r|\alpha, \beta)$ where Γ^{inv} is the inverse gamma distribution.

The Cox–Ingersoll–Ross (CIR) process (Cox et al. 1985) ensures that the mean of θ reverts to a $\Gamma(\alpha, \beta)$ distribution, but can divert over time. The speed at which the rate reverts is determined by κ. The correlation between time 0 and time t is $e^{-\kappa t}$, so with the CIR prior the distribution for rates is not uniform throughout the tree.

When selecting 'uniform' we assume the rate is uniformly distributed in the range $0 \ldots 10\,000$, which means a large proportion of the prior indicates a large value, with a mean of 5000.

8.4.2   Running SNAPP

A SNAPP analysis is run just as any BEAST analysis by using the BEAST program with the XML file. Note that SNAPP is very computationally intensive and there is no GPU support (as in BEAGLE) yet. However, SNAPP is multi-threaded, and using more threads can have a dramatic impact on the run time. Experimentation is required to determine the optimal number of threads since performance depends on the data. It can appear that SNAPP hangs if the number of lineages is very large. To get a feeling for your data and how much SNAPP can handle, it is recommended to run the analysis with a small subset of the sequences, say not more than 20 for diploid and 40 for haploid data.

One important consideration when interpreting the population sizes estimated by SNAPP is that these are effective population sizes over the entire length of the branch, not census (absolute) population sizes. There are many phenomena which can cause a large discrepancy between effective and census populations sizes. A population bottleneck, for example, can significantly reduce the effective population size for the entire branch. In the other direction, geographic structure within a population increases the effective population size.

To understand these effects, Figure 8.5 shows a few issues with population size estimation. When lineages are very close together and all coalescent events take place in the lowest branches ending in the taxa, at the top of these branches most probably just a single lineage remains. In this situation, the population sizes for the lower branches can be accurate, but since at most one coalescent event can take place in all the other branches all other estimates are necessarily low with high uncertainty. Likewise, if the

Figure 8.5 Population size estimate issues for species tree (in grey) with three species and a single gene tree in black. Left, sequences too close, middle sequences too diverse, right insufficient sequences.

sequences are too diverse, all coalescent events happen in the root and no accurate population size estimates can be guaranteed for all lower-lying branches. Make sure you have at least a couple of sequences per species, otherwise it becomes impossible to get an estimate of population sizes. This is because with just a single lineage for a species, there will be no coalescent event in the branch of the species tree ending in the taxon for that lineage, and population size estimates rely on coalescent events.

It is a good idea to run a SNAPP analysis with different settings for rate priors to detect that the estimates are not just samples from the prior but are informed by the sequence data. If the rate estimates are shaped according to the prior parameters one must be suspect of the accuracy. You can see this by inspecting the log file in Tracer. When the estimate changes with the prior parameters in individual runs with different priors you know that the estimates are not informed by sequence data, and these estimates should be considered uninformative. In general, it is quite hard to get good population size estimates. Like with *BEAST analysis, the estimate for the species tree topology tends to be most accurate, while timing estimates tend to be less accurate, and population size estimates even less accurate.

If you have large population size estimates relative to estimates for most branches, this may indicate errors in the data. Incorrect species assignment for a sequence will create the appearance of high diversity within the population it is placed in. Existence of cryptic species in your data also leads to large effective population size estimates. The models underlying SNAPP, as well as most of the methods in BEAST, assume a randomly mating population, at least approximately. Violations of this assumption could have unpredictable impacts on the remaining inferences. If cryptic species or population substructure is suspected it is preferable to rerun the analysis with subpopulations separated. In summary, large population size estimates should be reason for caution. By using model selection it is possible to reliably assign lineages to species and perform species delimitation (Leaché et al. 2014).

SNAPP is currently used in analysis of human SNPs, including some based on ancient DNA sequences, European and African blue tits, the latter in Morocco and on the Canary Islands, and various species of reptiles and amphibians such as western fence lizards (*Sceloporus occidentalis*), horned lizards (*Phrynosoma*), West African forest geckos (*Hemidactylus fasciatus* complex), African agama lizards (*Agama agama* complex), tailed frogs (*Ascaphus truei*) and African leaf litter frogs (*Arthroleptis poecilonotus*) (personal communications).

9 Advanced analysis

9.1 Sampling from the prior

There are a number of reasons to run an analysis in which you sample only from the prior before running the full analysis. One reason to do this is to confirm that a prior is proper. Here we mean proper in a strict mathematical sense, that is, that the prior integrates to unity (or any finite constant).

Although there are some situations in which improper priors can be argued for (Berger and Bernardo 1992), having an improper prior typically results in an improper posterior and if that is the case then the results of the MCMC analysis will be meaningless, and its statistical properties undefined. Often the observed behaviour of the chain will be that some parameters meander to either very large or very small values and never converge to a steady-state target distribution. For example, a uniform prior with bounds 0 and $+\infty$ for a molecular clock rate will result in the clock rate wandering to extreme values when an attempt is made to sample the prior in the absence of data. Even if no problems are evident when sampling the posterior, certain analyses rely on a proper prior and will return invalid results regardless of whether the posterior appears to be sampled correctly (an example of a method that requires a proper prior is path sampling for model comparison, see Section 1.5.7). For this reason, you should chose proper priors unless you know what you are doing.

Another reason to sample from the prior is to make sure that the various priors do not produce an unexpected joint prior in combination, or if they do, to check that the resulting prior is close enough to the practitioner's intentions. Especially when calibrations are used this can be an issue (Heled and Drummond 2012, 2013) since calibrations are priors on a part of a tree and a prior for the full tree is usually also specified (like Yule or coalescent). This means there are overlapping priors on the same parameter, which can produce unexpected results. Another situation occurs where a calibration with an upper bound on an ancestral clade sets an upper bound on the age of all the descendant clades, since none of them can exceed the age of their direct ancestor. When a non-bounded calibration is specified on a descendant this prior will be truncated at the top by the calibration on the ancestral clade. This phenomenon arises because calibrations are specified as one-dimensional marginal priors, even though it would be more appropriate to directly specify multi-dimensional priors when more than one divergence is calibrated. The result of specifying independent one-dimensional prior on divergences that are mutually constrained (e.g. $x < y$) can, for example, show up

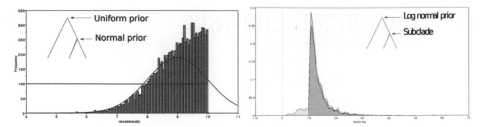

Figure 9.1 Left, a distribution of a clade by sampling from the prior. The clade has a normal (mean = 9, σ = 1 indicated by red line) calibration and a superclade has a uniform distribution with upper bound of 10 (blue line). The distribution still has a mean close to 9, but is now clearly asymmetric and the probability mass of the tail on the right is squeezed in between 9 and 10. Right, a marginal density of MRCA time of a clade with a log-normal prior together with the distribution of the MRCA time of a subclade and of a superclade. Note that the subclade follows the parent's distribution mixed with tail to the left. The superclade has almost the same distribution as the calibrated clade since the Yule prior on the tree tries to reduce the height of the tree, but the calibration prevents it from pushing it lower.

as a truncated distribution when sampling from the prior (see Figure 9.1). If there is any reason to believe a clade for which there is a calibration is monophyletic, it is always a good idea to incorporate this information as a monophyletic constraint since this typically reduces the complexity of interaction between multiple calibrations and simplifies quantification of the joint prior.

The easiest way to sample from the prior is to set up an analysis in BEAUti and on the MCMC panel click the 'Sample from prior' checkbox. Running such an analysis is typically very fast since most of the computational time in a full analysis is spent calculating the phylogenetic likelihood(s) (see Section 3.7 for details of the algorithm), which are not computed when sampling from the prior. After verifying that the prior distribution is an adequate representation of your prior belief, uncheck the checkbox in BEAUti and run the full analysis.

9.2   Serially sampled data

Calibrating one or more internal nodes (*node dating*), as described in Section 4.4, is one way to introduce temporal information into an analysis. Another method to achieve this is to use the sampling dates of the taxa themselves. Especially with fast-evolving species such as RNA viruses like HIV and influenza, the evolutionary rate is high enough and the sampling times wide enough that rates can be estimated accurately (Drummond et al. 2003a).

Note that having serially sampled data is not always sufficient to establish a time scale for the tree, or equivalently estimate molecular clock rates with any degree of certainty. Figure 9.2 illustrates why in some situations having dated tips is not sufficient to get an accurate divergence-time estimate. If the age of the root (t_{root}) is orders of magnitude greater than the difference between the oldest and most recent samples (the sampling interval Δt), one must wonder how accurate an estimate of the root age will be.

Figure 9.2 Left, tree without sampling dates for which no divergence-time estimate is possible without calibrations on internal nodes or a strong prior on the rate of the molecular clock. Middle, tree where tip date information is sufficiently strong (provided long sequences) to allow divergence-time estimates with a good degree of certainty, since the sample dates cover a relatively large fraction of the total age of the tree (sample interval $\Delta t = \frac{1}{3} t_{root}$). Right, tree where tip date information is available, but may not be sufficient for accurate estimation (sample interval $\Delta t = \frac{1}{25} t_{root}$), so divergence-time estimates will also be informed by the prior to a large extent.

Another consideration is the average number of substitutions along a branch spanning the sampling interval. This will be a product of the time Δt, the evolutionary rate k and the sequence length L. In cases where the information available from sampling times may not be sufficient, other calibration information, for example from the fossil record or from an informative prior on the molecular clock rate based on independent evolutionary rate estimates, may be used in combination with sampling date information to achieve accurate divergence-time estimates.

Some tree priors such as the Yule prior are only valid for contemporaneously sampled data, so not all tree priors can be employed when tip dates are introduced.

BEAUti provides support for entering taxa dates in the Tip Dates panel. When enabling tip dates, a table appears and you can edit the dates of sequences manually by simply double-clicking the appropriate cell in the table and typing in the value. Often, information about the sample dates is encoded in the taxa names. Instead of manually entering the dates, a utility is available that can extract the tip date from the taxon name. Also, if the tip dates data are stored in a file, such as a comma-separated file exported from a spreadsheet, you can import the dates using the 'guess dates' dialogue.

The *tip dating* method has been applied to ancient DNA to reconstruct a detailed prehistoric population history of bison (Shapiro et al. 2004), estimate the rate of evolution in Adélie penguins (Lambert et al. 2002) and resolve the taxonomy of the ratite moa (Bunce et al. 2003). Another large area of application is with fast-evolving pathogens, for example, the epidemiology of swine flu (Lemey et al. 2009b; Vijaykrishna et al. 2011), cholera (Mutreja et al. 2011) and simian immunodeficiency virus (SIV) (Worobey et al. 2010).

9.3 Demographic reconstruction

Relative population sizes through time can be estimated from a time-tree using coalescent theory as explained in Section 2.3. One of the nice features of such a reconstruction is that population bottlenecks can be detected. Furthermore, in epidemics, population size plots can be effective in detecting when epidemics started and whether policy changes for managing an epidemic have been effective (see Section 2.3.3 and Chapter 5 for models specifically designed for reconstructing epidemic population dynamics). To be able to reconstruct the population history you need to run an analysis with one of the coalescent tree priors, for example a coalescent prior with a parameterised population function such as constant or exponential growth, or (extended) Bayesian skyline plot (Drummond et al. 2005; Heled and Drummond 2008). Birth–death prior-based demographic plots are also possible (Kühnert et al. 2014; Stadler et al. 2013). There is support in Tracer (Rambaut and Drummond 2014) to reconstruct demographic histories, both from parametric coalescent analysis and Bayesian skyline plots (BSPs).

A demographic reconstruction appears as a graph showing population size history where the median (or mean) population size (actually $N_e\tau$) and 95% HPD intervals are plotted through time. An example is shown in Figure 9.3. Note that the y-axis does not represent absolute population size but a scaled population size, which is the product $N_e\tau$ where N_e is the effective population size and τ generation time (expressed in the same units as the divergence times). You might want to rescale the skyline plot and express it in N_e by dividing the population size parameter by generation time τ. The uncorrected

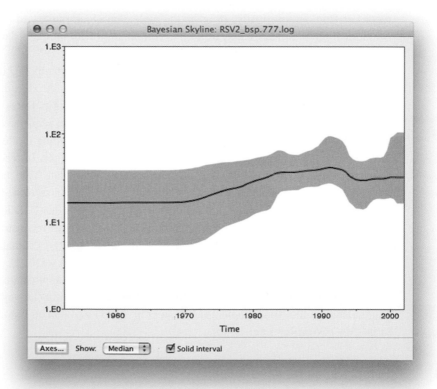

Figure 9.3 Bayesian skyline reconstruction in Tracer.

population size parameter values can be exported to a spreadsheet where they can be divided by τ and a new graph created. The median tends to be a more robust estimator when the posterior distribution has a long tail, and a log scale for the y-axis can help visualisation when population sizes vary over orders of magnitude, as can be the case during periods of exponential growth or decline.

The advantage of the extended Bayesian skyline plot (EBSP; Heled and Drummond 2008) over the BSP is that it does not require specifying the number of pieces in the piecewise population function. EBSP estimates the number of population size changes directly from the data using Bayesian stochastic variable selection (BSVS).

For EBSP analysis, a separate log file is generated, which can be processed with the EBSPAnalyser, which is part of BEAST 2. The program reads in the log file and generates a tab-delimited file containing time and population size mean, median and 95% HPD interval information that can be visualised with any statistical analysis package (e.g. R) or spreadsheet program.

The number of groups used by the EBSP analysis is recorded in the trace log file. It is not uncommon when comparing a BSP analysis with an EBSP analysis to find that (with default priors) the EBSP analysis uses on average fewer groups when analysing a single locus. As a result, the demographic reconstruction based on EBSP does not show a lot of detail, and in fact may often converge on a single group (which makes it equivalent to a constant population size model). Be aware that the default prior on the number of change points in the population size function may be prone to under-fitting, so that increasing the prior on the number of population size changes may increase the sensitivity of the analysis for the detection of more subtle signals in the data.

To increase the resolution of the demographic reconstruction, it is tempting to just add more taxa. However, due to the nature of coalescence, EBSP, like all coalescent methods, will generally benefit more from using multiple loci than from just adding taxa (Heled and Drummond 2008). So, adding an extra locus, with an independent gene tree but sharing the EBSP tree prior gives an increase in accuracy of the population size estimation that is generally larger than just doubling the number of taxa of a single alignment. This does not mean that adding more taxa does not help, just that adding another locus will generally help more than adding the same number of new sequences to existing loci. Likewise, increasing the length of the sequence tends not to help as much as increasing the number of loci. For a more refined analysis (Felsenstein 2006) has made a careful study of the sampling tradeoffs in population size estimation using the coalescent.

9.3.1 Some demographic reconstruction issues to be aware of

A BSP or EBSP reconstruction typically shows 95% HPD intervals that (going back into the past) can remain constant in both estimated population size and its uncertainty for a large portion at the end of the plot. Intuitively, one would expect the uncertainty in the estimate to increase going back in time since the further one goes back in time the more distant we are from the sampled data (when there is no calibration information). This holds especially for the the oldest part of the plot.

However, the reason these HPD intervals do not behave as expected is that the boundaries of the underlying piecewise-constant population function coincide with coalescent events (by design). If instead the boundaries were chosen to be equidistant, then the expected increase in uncertainty going back towards the root would occur because on average there would be fewer coalescent events in intervals that are nearer to the root *and* the estimated times of those coalescent events nearer the root would also be more uncertain. However, because of the designed coincidence of population size changes at coalescent events in the BSP and EBSP, the number of coalescent events per interval (i.e. the amount of information available to estimate population size per interval) is *a priori* equal across intervals. So in (E)BSP models, uncertainty in population size is traded for a loss of resolution in the timing of population size changes (since the underlying population size can typically only change at wide intervals near the root of the tree).

Note that this modelling choice is different from that used in the birth–death skyline plot (Stadler et al. 2013), in which the piecewise function is chosen to have changes at regularly spaced intervals. Consequently, (net) birth rates reconstructed through time using the birth–death skyline model tend to show 95% HDP intervals that grow larger the further one goes back in time.

It is sometimes the case that a horizontal line can be drawn through the 95% HPD intervals of the resulting BSP. This doesn't mean that the reconstruction is consistent with a constant population size. A better way to decide whether there is a trend in population size is to count the fraction of trees in the posterior sample that suggest a trend. For example, if more than 95% of the posterior samples describe a skyline plot where the population size at the root is smaller than at the tips, then a growing population size can be inferred. In the case of EBSP, a more direct test is simply to calculate the posterior probability that there is more than one piece in the population size function.

One of the reasons that a demographic reconstruction can come with large 95% HPD intervals suggesting large uncertainty in population sizes is due to uncertainty in the molecular clock rate. Figure 9.4 shows the effect of having low uncertainty in the relative divergence times, but large uncertainty in the overall molecular clock rate. If only trees corresponding to low molecular clock rates and thus old trees were considered in the posterior, the BSP would exhibit narrow 95% HPD intervals; likewise for trees associated with high molecular clock rates. However, since the posterior contains both the old and young trees, together the 95% HPD interval spans the range and the HPD bounds become quite large. A BSP drawn in units of fractions of the root age or based on a fixed clock rate (say the posterior median estimate of the clock rate) would show much smaller credible intervals, and may uncover a strong underlying signal for population dynamics, that would otherwise be confounded.

Finally, one has to be aware that non-parametric methods like BSP can contain systematic bias due to sampling strategies when using measurable evolving populations (Silva et al. 2012). This can lead to conclusions that an epidemic slowed, while the facts do not support this. On the other hand, parametric methods can give unbiased estimates if the population sizes are large enough, and the correct parametric model is employed.

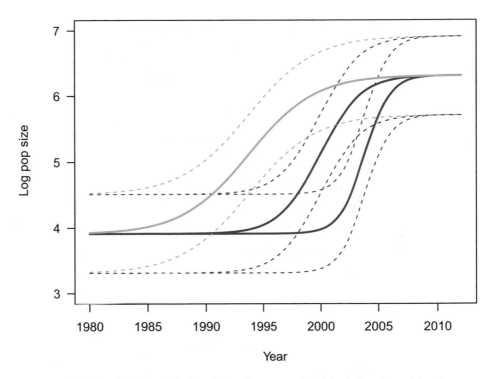

Figure 9.4 A BSP analysis in which the relative divergence times are well estimated, but the absolute times are not, will produce a posterior distribution containing trees with a wide variation of root ages. It may be the case that conditional on a particular root age, there is a strong trend in population size through time with low uncertainty, but as a result of the uncertainty in the absolute time scale the overall result is a BSP with large uncertainty. In this hypothetical example, the time-trees with old root ages have a BSP indicated by the curve to the flattest curve (green) with (conditional) 95% HPD indicated by dashed lines. Likewise, the youngest time-trees from the posterior sample can have a BSP indicated by the curve with the youngest transition in population size (blue). The middle curve shows a conditional demographic estimate associated with an average tree age. However, together the 95% HPD will cover all of these curves, which shows up as large 95% HPD intervals in the skyline plot of the full posterior.

Applications: Some notable applications of demographic reconstruction are determining the pandemic potential of the influenza H1N1 strain (Fraser et al. 2009), and tracking feline immunodeficiency virus in order to reveal population structure of its host, *Puma concolor* (Biek et al. 2006). It is possible to detect population bottlenecks, for example in Moroccan sardines (Atarhouch et al. 2006), fur seal (*Arctocephalus gazella*) (Hoffman et al. 2011) and hydrothermal shrimp (Teixeira et al. 2012), to name just a few. It was used to demonstrate population decline in African buffalo during the mid-Holocene (Finlay et al. 2007) and ancient bison (Drummond et al. 2005), as well as population expansions in domesticated bovines (Finlay et al. 2007) and lack of population decline in woolly mammoth during a long period before its demise (Debruyne et al. 2008).

9.4   Ancestral reconstruction and phylogeography

BEAST can perform various phylogeographical analyses. One way to look at these analyses is as extending alignments with extra information to indicate the location. For discrete phylogeographical analysis, a single character (data column) is added and for continuous analysis a latitude–longitude pair is added. These characters share the tree with the alignment, but have their own substitution and clock model. That is why for discrete and continuous phylogeography they are treated in BEAUti as just a separate partition, and they are listed in the alignment tab just like the other partitions.

Discrete phylogeography (Lemey et al. 2009a) can be interpreted as a way to do ancestral reconstruction on a single character, which represents the location of the taxa. Sampled taxa are associated with locations and the ancestral states of the internal nodes in a tree can be reconstructed from the taxon locations. In certain circumstances it is not necessarily obvious how to assign taxa to a set of discrete locations, or indeed how to choose the number of discrete regions used to describe the geographical distribution of the taxa. The number of taxa must be much larger than the number of regions for the analysis to have any power. This approach was applied to reconstruction of the initial spread of human influenza A in the 2009 epidemic (Lemey et al. 2009b), tracing *Clostridium difficile*, which is the leading cause of antibiotic-associated diarrhoea worldwide (He et al. 2013) and geospatial analysis of musk ox with ancient DNA (Campos et al. 2010).

Besides the ancestral locations, the key object of inference in a phylogeographical model is the migration rate matrix (see Chapter 5). In a symmetrical migration model with K locations there are $\frac{K(K-1)}{2}$ migration rates, labelled $1, 2, \ldots, \frac{K(K-1)}{2}$. These labels are in row-major order, representing the rates in the upper right triangle of the migration matrix (see Figure 9.5). For a non-symmetric migration model the number of rates is $K(K-1)$.

For continuous phylogeography (Lemey et al. 2010), the locations (or regions) of the individual taxa need to be encoded in latitude and longitude. A model of migration via a random walk is assumed, and this makes such an analysis a lot more powerful than a discrete phylogeography analysis since distances between locations are taken into account. For many species, a random walk is a reasonable null model for migration, especially when the dispersal kernel is allowed to be over-dispersed compared to regular Brownian motion (Lemey et al. 2010). However, for species that have seasonal migrations such as shore birds, this may not be a good null model. Likewise, viruses transmitted by humans

	A	B	C	D	E			A	B	C	D	E
A	-	1	2	3	4		A	-	1	2	3	4
B	1	-	5	6	7		B	5	-	6	7	8
C	2	5	-	8	9		C	9	10	-	11	12
D	3	6	8	-	10		D	13	14	15	-	16
E	4	7	9	10	-		E	17	18	19	20	-

Figure 9.5 Left, a matrix for ancestral reconstruction with five locations {A,B,C,D,E}, with symmetric rates. Right, a matrix with five locations {A,B,C,D,E}, with non-symmetric rates.

migrate efficiently through aeroplane travellers, suggesting approaches for modelling their geospatial spread that directly incorporate transportation data (see Lemey et al. 2014).

Visualisation of the posterior estimates from a continuous phylogeographic analysis can be achieved by using SPREAD (Bielejec et al. 2011), which produces output for Google Earth. This continuous phylogeographic model has been used to reconstruct the origin of Indoeuropean languages (Bouckaert et al. 2012), perform an analysis of the spread of HIV in Cameroon (Faria et al. 2012) and determine the origin of maize streak virus (Monjane et al. 2011). Recently, a link between phylogenetics and mathematical spatial ecology was established (Pybus et al. 2012), extending the relaxed random walk into analysis of epidemics.

The simplest versions of continuous models are not aware of landscape features so cannot distinguish between different rates of migration, for example over land versus over water. The landscape-aware model (Bouckaert et al. 2012) can distinguish between such features by rasterising the landscape into a digital feature map. In fact it converts the problem into a discrete phylogeographical analysis where the substitution model is created based on the underlying landscape through a rather computationally intensive pre-calculation. The landscape-aware model will have more power than the discrete model since it takes distances into account, just as the continuous model. There will be only minor differences between the continuous and landscape-aware model if the landscape has rather uniform rates. However, when rates between different landscape features vary, for example when the rate of migration along shorelines is assumed to be much higher than migration over land or on water, then the landscape-aware model can produce large differences in the phylogeographical reconstruction. Also, the landscape-aware model makes it relatively easy to reconstruct the path along a branch, as shown in Figure 9.6.

For discrete and continuous phylogeographical analysis you need the `beast-classic` package and the landscape-aware model is implemented in the `beast-geo` package.

Testing phylogeographic hypotheses: For a discrete phylogeographical analysis (similarly as for a discrete trait), you might want to calculate the support for the geographic location of a certain common ancestor or for the root of the tree. The proportion of the sampled trees that fit the hypothesis gives the posterior probability that the hypothesis is true. TreeAnnotator creates a summary tree and labels every internal node with the distribution of the geographical locations at that node in the tree. You can use FigTree or IcyTree to inspect these distributions. Note that it is important to compare with the prior distribution for the locations at the clades of interest to ensure that the posterior distribution is informed by the data instead of just reflecting the prior.

9.5   Bayesian model comparison

Since BEAST provides a large number of models to choose from, an obvious question is which one to choose. The most sound theoretical framework for comparing two models

Figure 9.6 Reconstruction of hepatitis B migration from Asia through northern North America using the landscape-aware model. The summary tree is projected onto the map as a thick line, and the set of trees representing the posterior is projected onto the map as lightly coloured dots indicating some uncertainty in the migration path, especially over some of the islands. Note the migration backwards into Alaska after first moving deeply into Canada. The model assumed much higher migration along coastlines than over water or land.

or hypotheses, M_1 and M_2, in a Bayesian framework is to calculate the Bayes factor (BF; see Section 1.5.6).

Let us consider a hypothetical example where model M_1 is a BSP tree prior with strict molecular clock and model M_2 is a constant-sized coalescent tree prior and a relaxed molecular clock. We want to decide which model best explains the data. Before embarking on a complex model comparison analysis using BFs, as a rule of thumb, if the marginal posterior distributions of the log-likelihoods from the two models do not overlap then it is quite safe to assume the model with higher posterior distribution of log-likelihoods is preferred. Note that we are comparing the marginal posterior distributions of the *likelihoods*, not the log posterior statistic itself. The log posterior is not generally comparable between models because not all priors will necessarily include their normalising constants. Likelihoods, on the other hand, are guaranteed to sum to unity when adding over all possible data sets, so likelihoods of quite different models (such as model M_1 and M_2 above) are directly comparable. Sadly, the marginal posterior distributions of the likelihoods of competing models often do overlap and thus a more sophisticated model comparison approach is required. We elaborate on some options in the following sections.

9.5.1   Bayes factors

A number of methods for calculating BFs involve running one or more MCMC analyses for each model (e.g. Baele et al. 2012; Lartillot and Philippe 2006; Xie et al. 2011).

From a practical standpoint, it should be obvious that you cannot do a valid BF analysis if one of the chains has not converged, as indicated by low effective sample sizes (ESSs). So unless you are able to get good ESSs for all competing models, your BF estimates are meaningless. A low ESS for a particular model might indicate that there is not much signal in the data and you will have to make sure your priors on the model parameters are proper and sensible. Also, see troubleshooting tips (Section 10.3) for more suggestions.

Once you have adequate posterior samples from each of the competing models, the easiest, fastest but also least reliable way (Baele et al. 2012) to compare models is to use the harmonic mean estimator (HME; see Section 1.5.6) to estimate the marginal likelihood. This can be done in Tracer (Rambaut and Drummond 2014) by loading the log files and using the Analysis → Model Comparison menu. The HME is known to be very sensitive to outliers in the posterior sample. Tracer implements a smoothed HME (sHME) that employs a bootstrap approach (Redelings and Suchard 2005) by taking a number of pseudoreplicates from the posterior sample of likelihoods so that the standard deviation of the estimate can be calculated, which accounts for the effect of outliers somewhat. Still, even the sHME is not recommended since like the HME it tends to produce poor estimates of the marginal likelihood (Baele et al. 2013a).

Path sampling and the stepping stone algorithm (Baele et al. 2012; Xie et al. 2011) are more advanced algorithms for estimating the marginal likelihood (see Section 1.5.7). They work by running an MCMC chain that samples a family of target distributions $\pi_\beta(\theta) = f(\theta)\Pr(D|\theta)^\beta$ constructed from a schedule of values of β where $f(\theta)$ is the prior, $\Pr(D|\theta)$ the likelihood, θ the parameters of the model and D the data. When $\beta = 0$, the chain samples from the prior, when $\beta = 1$, it samples from the posterior and intermediate values of β bridge between these two extremes. Of the two approaches, the stepping stone method tends to be more robust. Empirically, a set of values for β, the steps, that gives an efficient estimate of the marginal likelihood (Xie et al. 2011) is obtained by following the proportions of a $\beta(0.3, 1.0)$ distribution. To set up a path sampling analysis in BEAST you need to set up a new XML file that refers to the MCMC analysis of the model. To reduce computation on burn-in, the end-state of a run for one value β is used as the starting state for the next β value. This works for every value of β, except for the first run, which requires getting through burn-in completely. So, you need to specify (1) a burn-in for the first run, (2) a burn-in for consecutive runs and (3) the chain length for generating the samples. Appropriate values depend on the kind of data and model being used, but burn-in for the first run should not be less than burn-in used for running a standard MCMC analysis on the same data. The log files can be inspected with Tracer to see whether the value of burn-in is sufficiently large and that the chain length produces ESSs such that the total ESS used for estimating the marginal likelihood is sufficiently large.

The number of steps needs to be specified as well, and this number is also dependent on the combination of model and data. To determine an appropriate number of steps, run the path sampling analysis with a low number of steps (say ten) first, then increase the number of steps (with, say, increments of ten, or doubling the number of steps) and see whether the marginal likelihood estimates remain unchanged. Large differences between estimates indicate that the number of steps is not sufficiently large. It may not be practical to run a path sampling analysis because of the computational time involved, especially for large analyses where running the main analysis can take days or longer. In these cases a pragmatic decision can be made using the AICM method instead (see Section 1.5.6 for details).

The efficiency of running a path sampling analysis can be improved by using threads or a high-performance cluster. Model comparison is an active area of research, so it is possible that new more efficient and more robust methods will be available in the near future.

9.6   Simulation studies

Simulation studies can be used to find out the limits of the power of some models. Some questions that can be answered using simulation studies are:

- How much sequence data is needed to estimate the model parameters reliably?
- How well can a tree topology be recovered using a specific model?
- How uncertain are rate estimates under various tip sampling schemes?

BEAST contains a sequence simulator that can be used to generate synthetic multiple sequence alignments. It requires specification of a tree-likelihood, with its tree, site model, substitution model and molecular clock model, and generates a random sequence alignment according to the specification. The tree can be a fixed tree specified in Newick, or a random tree generated from a coalescent process, possibly with monophyletic constraints and calibrations. The site and substitution model, as well as the molecular clock model, can be any of the ones available in BEAST or one of its packages. However, note that using a non-strict clock model requires extra care because of the way it is initialised.

To perform simulation studies that involve dynamics of discrete populations the Moments and Stochastic Trees from Event Reactions (MASTER) (Vaughan and Drummond 2013) package can be used. It supports simulation of single and multiple population size trajectories as well as corresponding realisations of the birth–death branching processes (trees). Some applications include: simulation of dynamics under a stochastic logistic model, estimating moments from an ensemble of realisations of an island migration model, simulating an infection transmission tree from an epidemic model and simulating structured coalescent trees. MASTER can be used in conjunction with a sequence simulator to generate alignments under these models. For example, a tree simulated from the structured coalescent can be used to generate an alignment, which can be used to try to recover the original tree and the population parameters that generated it.

10   Posterior analysis and post-processing

In this chapter we will have a look at interpreting the output of an MCMC analysis. At the end of a BEAST run, information is printed to the screen, and saved in trace log and tree log files. This chapter considers how to interpret the screen and trace log, while the next chapter deals with tree logs. We have a look at how to use the trace log to compare different models, and diagnose problems when a chain does not converge. As you will see, we emphasise comparing posterior samples with samples from the prior, since you want to be aware whether the outcome of your analysis is due to the data or a result of the priors used in the analysis.

Interpreting BEAST screen log output: At the end of a BEAST run, some information is printed to screen (see listing on page 88) detailing how well the operators performed. Next to each operator in the analysis, the performance summary shows the number of times an operator was selected, accepted and rejected. If the acceptance probability is low (< 0.1) or very low (< 0.01) this may be an indication that either the chain did not mix very well, or that the tuning parameters for the operator were not appropriate for this analysis. BEAST provides some suggestions to help with the latter case. Note that a low acceptance rate does not necessarily mean that the operator is not appropriately parameterised. For example, when the sequence alignment data strongly support one particular topology, operators that make large changes to the topology (like the wide exchange operator) will almost always be rejected. So, some common sense is required in interpreting low acceptance rates.

If the acceptance rate is high (> 0.5), this indicates that the operator is probably making jumps that are too small, and BEAST may produce a suggestion to change a parameter setting for the operator. The exception to this are operators that use the Gibbs distribution (Geman and Geman 1984), which are generally efficient and always accepted.

For relaxed clock models, if the uniform operator on the branch-rate categories parameter has a good acceptance probability (say > 0.1) then you do not need the random walk integer operator on branch-rate categories. You could just remove it completely and increase the weight on the uniform operator on branch-rate categories. Any operator that changes the branch-rate categories parameter changes the rates on the branches, and thus the branch lengths in substitutions and therefore the likelihood.

10.1   Trace log file interpretation

Probably the first thing to do after a BEAST run has finished is to determine that the chain has converged to the target distribution. There are many statistics for determining whether a chain has converged (Brooks and Gelman 1998; Gelman and Rubin 1992; Gelman et al. 2004; Geweke 1992; Heidelberger and Welch 1983; Raftery and Lewis 1992; Smith 2007), all of which have their advantages and disadvantages, in particular sensitivity to burn-in. The practitioner of MCMC should always visually inspect the trace log in Tracer (see Figure 6.7 for a screenshot) to detect obvious problems with the chain, and check the ESS (Kass et al. 1998).

How much burn-in? The first thing to determine is the amount of burn-in that the chain requires. Burn-in is the number of samples the chain takes to reach the target distribution equilibrium and these samples must be discarded. By default, Tracer assumes that 10% of the chain is burn-in and if this is not sufficient the chain probably needs to be run for longer. In fact, a hardline view held by some Bayesian MCMC aficionados is that if any burn-in is required, then the chain has not been run long enough. More pragmatically, you really need to visually inspect the graph of each of the posterior statistics that are logged since it is usually easy to see when burn-in is reached; during burn-in the values increase or decrease steadily and an upward or downward trend is obviously present. After burn-in the trace should not show a trend any more, and oscillate around a stationary distribution (the target distribution). The smaller the amount of burn-in, the more samples are available. If a substantial proportion of the chain ($>10\%$) is burn-in, it becomes difficult to be sure the remaining part is a good representation of the posterior distribution. In such a case it would be prudent to run the chain for, say, ten times longer, or run many independent chains and combine their post-burn-in portions.

As you can see in Figure 10.1, it is very important to actually visually inspect the trace output to make sure that burn-in is removed, and when running multiple chains that all the runs have converged on the same distribution.

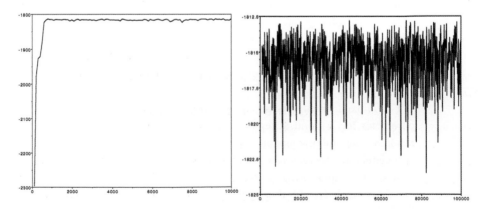

Figure 10.1 Left, trace with an ESS of 36 that should be rejected, not only due to the low ESS but mainly due to the obvious presence of burn-in that should be removed. Right, trace with ESS of 590 that is acceptable.

All about ESS: After determining the correct burn-in, the next thing to check is the ESS (Kass et al. 1998) of the sampled parameters. The ESS is the number of independent draws from the posterior distribution that the Markov chain is equivalent to. This differs from the actual number of samples recorded in the trace log, which are autocorrelated and thus dependent samples. In general, a larger ESS means the estimate is a better approximation of the true posterior distribution. Tracer flags ESSs smaller than 100, and also indicates whether ESSs are between 100 and 200. But this may be a bit liberal and ESSs over 200 are more desirable. On the other hand, chasing ESSs larger than 10 000 is almost certainly a waste of computational resources. Generally ESSs over 200 would be adequate for most purposes so $\approx$1000 is very good. If the ESS of a parameter is small then the estimate of the marginal posterior distribution of that parameter can be expected to be poor. In Tracer you can calculate the standard error of the estimated mean of a parameter (note that this is not the standard deviation of the parameter, but the error in the estimate of the mean of the marginal posterior distribution of the parameter). If the ESS is small then the standard error will be large, and vice versa.

The ESSs of parameters in the log file don't necessarily tell you whether the MCMC chain is mixing in phylogenetic tree space. At the moment BEAST does not include any tools for directly examining the ESS of the tree or clade statistics. For this purpose a program like AWTY can be used (Nylander et al. 2008).

Increasing ESS: These are some ways of increasing the ESS of a parameter:

- The most straightforward way of increasing the ESS is to increase the chain length. This obviously requires more computer resources and may not be practical.

- If only a few items get low ESSs, generally you want to reduce the weight on operators that affect parameters with very high ESSs and increase weights on operators operating on the parameters with low ESSs.

- You can increase the sampling frequency. The ESS is calculated by measuring the correlation between sampled states in the chain which are the entries in the log file. If the sampling frequency is very low these will be uncorrelated. This will be indicated by the ESS being approximately equal to the number of states sampled in the log file (minus the burn-in). If this is the case, then it may be possible to improve the ESSs simply by increasing the sampling frequency until the samples in the log file begin to be autocorrelated. However, be warned that sampling too frequently will not affect the ESSs but will increase the size of the log file and the time it takes to analyse it.

- Combine the results of multiple independent chains. It is a good idea to do multiple independent runs of your analyses and compare the results to check that the chains are converging and mixing adequately. If this is the case then each chain should be sampling from the same distribution and the results could be combined (having removed a suitable burn-in from each). The continuous parameters in the log file can be analysed and combined using Tracer. The tree files will currently have to be combined manually using a text editor or LogCombiner. An advantage of this approach is that the different runs can be performed simultaneously on different computers (say, in a computer lab or nodes on a cluster) or on different processors in a multi-processor machine.

When to combine log files: A common question is whether it is acceptable to use an estimate of, say, MRCA time, from the combined analysis of three identical runs (from different random seeds) if the ESS is >200 for all parameters in the combined file (but not in each individual file) in Tracer.

Three runs combined to give an ESS of >200 is probably safe but you need to be a little bit careful. Ten runs to get an ESS of >200 might not be safe because if individual runs have ESS estimates of about 20 then there is question of whether those ESS estimates are valid at all. You should visually inspect that your runs are giving similar answers for all the parameters as well as the likelihood and prior. The traces should substantially overlap, and be giving the same mean and variance (i.e. within two standard errors). If your runs give traces whose target distributions do not overlap then you cannot combine them, regardless of the magnitude of the combined ESS.

When Tracer combines two traces it just reports the sum of the ESS of the two runs. However, if you combine the two runs using `LogCombiner` and then look at the ESS in Tracer then Tracer will treat the combined run as a single run and calculate the ESS based on the autocorrelation. If, for example, the two runs have each got stuck on a different local optimum with different distributions, then Tracer will detect that the first half of the concatenated trace is different from the second half and will report a low ESS. However, if you combined these two within Tracer, it would just add the ESSs and everything would appear at first glance to be fine. Note that when combining logs in LogCombiner, if not enough burn-in is removed this may result in poor estimates of ESS. Make sure the combined log looks like all samples are from the same distribution to make sure that the combined log does not contain burn-in. Burn-in that is not removed will show up as discontinuities in the trace plot.

Marginal posterior distribution: Tracer can plot a density diagram showing the marginal posterior distribution in a graph. When inspecting the marginal posterior distribution of a parameter of interest, also have a look at the prior on the parameter. The difference between the two shows how much the data informs the parameter estimate. The comparison between the two is easier when the priors are carefully chosen. Preferably, use proper priors and not so diffuse that it is hard to estimate the prior density at parameter values of interest.

Typically, for most scale parameters in a phylogenetic inference the marginal posterior distribution will be unimodal.

Mean and its standard error: Another useful set of statistics provided by Tracer is the mean and standard error of the mean. The standard error of the sample mean is defined as the standard deviation of the sample divided by the square root of the sample size. Note that instead of using the actual number of samples, the ESS is used as the sample size since the samples are correlated and the ESS is an estimate of the sample size that corrects for this autocorrelation. If the ESS is small then the standard error will be large, and vice versa. Also, note that the standard error applies only to the mean and does not say anything about how well the standard deviation of the posterior is estimated.

x% **HPD interval:** Tracer shows the 95% highest posterior density (HPD) interval of every item that is in the log. In general, the *x*% HPD interval, is the *smallest* interval that contains *x*% of the samples. If one tail is much longer than the other then most of the removed values will come from the longer tail.

To calculate the *x*% HPD interval, sort the values and consider the interval starting from the very smallest value to the *x*-percentile value and store the difference in the values (that is, the size of the interval). Now increment one value at a time (at both ends, so that you always have *x*% of the values in the interval) until you get to the interval that starts at the (100-*x*)-percentile value and finishes at the largest value. Each step compares the new interval width to the minimum found so far. If it is smaller than the minimum so far, update the minimum and store the start value. This algorithm takes $O(N)$ operations where N is the size of the sample. Since the sample needs to be sorted first, which takes $O(N \log N)$, this dominates the calculation time.

The *x*% central posterior density (CPD) interval, on the other hand, is the interval containing *x*% of the samples after $[(100\text{-}x)/2]\%$ of the samples are removed from each tail. The shorthand for both types of interval is the *x*% *credible interval*.

Clock rate units: Tracer shows the mean of the posterior but not the units. The units for the clock rate depends on the units used for calibrations. When a calibration with, say, a mean of 20 is used representing a divergence date 20 million years ago, the clock rate is in substitutions per site per million years. If all digits are used, that is 20 000 000 or 20E6 for the mean of the calibration, the clock rate unit will be substitutions/site/year. If the calibration represents 20 years ago, the unit is substitutions/site/year. Likewise, units for tree divergence ages are dependent on the units used to express timing information; with a calibration of 20 representing 20 million years ago, a divergence age of 40 means 40 million years ago.

If there is no timing information, by default the clock rate is not estimated and fixed to 1, and branch lengths represent substitutions per site. Of course, if the clock rate is estimated a prior on the clock rate needs to be provided and the units for clock rate is equal to the units used for the clock rate prior. Alternatively, the clock rate can be fixed to a value other than 1 if an estimate is available from the literature or from an independent data set. The ages of nodes in the tree are in units of time used for the clock rate. If the clock rate is expressed in substitutions/site/million years then estimated divergence ages will be expressed in million of years.

Age of tree and divergence times: To interpret the age of the tree (root height) and divergence times in Tracer, you have to keep in mind that time runs backwards in BEAST by default. The most recently sampled sequence has an age of zero, and the age of the tree, at say 340, represents the MRCA of all sequences, at 340 years (assuming your units are in years) in the past. If you have a taxon with two virus samples, one from the year 2000 and one from the year 2005, and you get an MRCA time value of 10, these viruses diverged in 1995. Note that MRCA time is counted from the youngest taxon member.

Coefficient of variation: With relaxed clock models a coefficient of variation (c_v) is logged, which is defined as the standard deviation divided by the mean of the clock rate. This is a normalised measure of dispersion. The coefficient gives information about how clock-like the data are. A coefficient of 0.633 means that it was estimated that the difference in the rate of evolution of two typical lineages in the analysis varied by 63.3% of the absolute clock rate. Values closer to zero indicate the data are more clock-like and a strict clock may be more appropriate. There is no strict rule, but values below 0.1 are generally considered to be low enough to justify the use of a strict molecular clock model (e.g. see Brown and Yang 2011). For the log-normal distribution c_v is a simple function of the standard deviation of the log rate S: $c_v = \sqrt{\exp(S^2) - 1}$. For $S \ll 1$, $c_v \approx S$.

If the marginal posterior distribution of the S or c_v extends to values very close to zero, this is also an indication that the strict clock model cannot be rejected. Using a strict molecular clock for clock-like data has the advantage of requiring fewer parameters and increases precision of rate estimates (Ho et al. 2005) and topological inference (Drummond et al. 2006) without compromising accuracy. If the value of the coefficient of variation is large (e.g. $c_v \gg 0.1$) then the standard deviation is larger and a relaxed molecular clock is required (see Figure 10.2). If S or c_v is greater than 1, then the data are very non-clock-like and probably shouldn't be used for estimating divergence times.

A relaxed molecular clock with exponential distribution of rates across branches will always have a coefficient of variation of 1.0. It is a one-parameter distribution and the parameter determines both the mean and the variance. The only reason that BEAST reports a number slightly under 1 is because of the way the distribution is discretised across the branches in the tree. So you should ignore the ESS estimates for this statistic when using the exponential model. If you do not think that the coefficient of variation

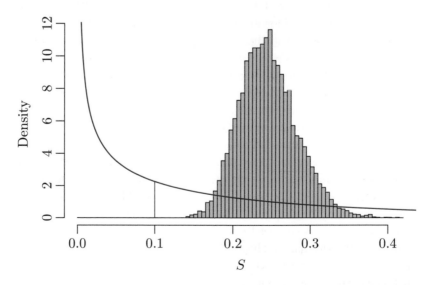

Figure 10.2 The gamma-distributed prior and the posterior histogram for the S parameter of the uncorrelated log-normal relaxed molecular clock from an analysis of the *Anolis* data set. The prior median of S is 0.1 and the posterior probability of $S > 0.1$ is 100%, providing strong evidence against a strict molecular clock.

is about 1 for your data, then you probably should not use the exponentially distributed rates across branches model.

Rates for clades: Rates for clades are not logged in the trace log but can be found in the tree log. You can calculate a summary tree with TreeAnnotator and visualise the rates for a clade in FigTree or IcyTree.

10.2   Model selection

In Sections 1.5.6 and 9.5 we have already looked at comparing models using Bayes factors (BFs). Here we look at strategies for comparing the various substitution models, clock models and tree priors.

Model selection strategy: The first rule to take into account when selecting a model is not to use a complex model until you can get a simple model working. A good starting point is the HKY substitution model, with a strict molecular clock and a constant population size coalescent tree prior (if you want a coalescent tree prior, otherwise start with Yule instead). If you cannot get convergence to the target distribution using this simple model then there is no point in going further. Convergence means running several independent chains long enough that they give the same answers and have large enough ESSs. If you can get this working then add the parameters you are interested in. For example, if you want to get some information about population size change, try BSP with a small number of groups (say four) or EBSP, which finds the number of groups. You should probably add gamma rate heterogeneity across sites to your HKY model once everything else is working.

The important thing about model choice is the sensitivity of the estimated *parameter of interest* to changes in the model and prior. So in many respects it is more important to identify which aspects of the modelling have an impact on the *answer you care about* than to find the 'right' model.

You could argue that a 'correct' analysis would compare all possible combinations of demographic, clock and substitution models, assuming your state of knowledge beforehand was completely naive about the relative appropriateness of the different models. This is a principled but extreme perspective. Nevertheless, there has been some progress in implementing model averaging over all the substitution models automatically within a single MCMC analysis. For example, the model from (Wu et al. 2013) is implemented in the `subst-bma` package and simultaneously estimates the appropriate substitution models and partitioning of the alignment. The reversible-jump-based model (Bouckaert et al. 2013) is a substitution model that jumps between models in a hierarchy of models, and is available through the RBS package as the RB substitution model. It can also automatically partition the alignment, but unlike `subst-bma` assumes that partitions consist of a fixed number of contiguous runs of sites. One can average over clock models in a single MCMC analysis as well (Li and Drummond 2012).

For the standard models (HKY, GTR, gamma categories, invariant sites) using Model-Test (Posada and Crandall 1998) or jModelTest (Posada 2008) is probably a reasonable thing to do if you feel that time is a precious commodity. However, some practitioners would assert that mixing ML and Bayesian techniques is not a principled approach, and ideally one would want to select or average over models in a Bayesian framework, if Bayesian inference is the final aim.

Regardless, if you have protein-coding sequences then you should consider a codon-position model by splitting the alignment into codon positions (Bofkin and Goldman 2007; Shapiro et al. 2006) and based on a BF decide between it and alternative substitution models.

One pleasant aspect of Bayesian inference is that it is easier to take into account the full range of sources of uncertainty when making a decision, like choosing a substitution model. With a BF you will be taking into account the uncertainty in the tree topology in your assessment of different substitution models, whereas in ModelTest you have to assume a specific tree (usually a neighbour-joining tree). While this may not be too important Yang et al. (1995) argue that if the tree is reasonable then substitution model estimation will be robust), the Bayesian method is definitely more satisfying in this regard.

A disadvantage of ModelTest for protein-coding sequences is that it does not consider the best biological models, those that take into account the genetic code, either through a full codon model (generally too slow computationally to estimate trees) or by partitioning the data into codon positions. See (Shapiro et al. 2006) for empirical evidence that codon position models are generally superior to GTR + Γ + I, and just as fast if not faster.

As far as demographic models are concerned, it is our experience that they do not generally have a great effect on the ranking of substitution models.

Strict vs. relaxed clock: The random local clock model (Drummond and Suchard 2010) can be used as a Bayesian test of the strict molecular clock. If the posterior probability of zero rate changes (i.e. a strict molecular clock) is in the 95% credible set, then the data can be considered compatible with a strict clock. Alternatively, a heuristic comparison between log-normal relaxed and strict molecular clocks is relatively easy. Use a log-normal relaxed clock first, with a prior on S that places 50% of the probability mass below $S = 0.1$ (e.g. a gamma (shape = 0.5396, scale = 0.3819) distribution has a median of ≈ 0.1 and has 97.5% of prior probability below $S = 1$). If the estimated $S > 0.1$ and there is no probability mass near zero in the marginal posterior distribution of S (see also Section 10.1) then you cannot use a strict clock. However, if the estimate of S is smaller than 0.1, then a strict clock can probably be safely employed (Brown and Yang 2011). Note that if the clock model is the parameter of interest, then you should employ formal model comparison, such as by calculating BFs (see Section 1.5.6).

Constant vs. exponential vs. logistic population: To select a population function for a coalescent tree prior, use exponential growth first. If the marginal posterior distribution of the growth rate includes zero, then your data are compatible with constant population size.

If you are only doing a test between exponential growth and constant then you can simply inspect the posterior distribution of the growth rate and determine if zero is contained in the 95% HPD. If it is, then you cannot reject a constant population size based on the data. Of course this is only valid if you are using a parameterisation that allows negative growth rates.

Note that if the population history is the parameter of interest, then you should employ formal model comparison, such as by calculating BFs using path sampling (see Sections 1.5.6 and 1.5.7).

Alternative *BEAST species trees: Suppose you have two alternative assignments of sequences to species in a *BEAST analysis, and you want to choose one of the species trees. If your observations were actually gene trees instead of sequence alignments then it would be straightforward: the species.coalescent likelihoods would be the appropriate likelihoods to be comparing by BFs. However, since we do not directly observe the gene trees, but only infer them from sequence alignments, the species.coalescent likelihoods are not likelihoods but rather parametric priors on gene trees, where we happen to care about the parameters and structure of the prior, that is, the species tree topology and the assignments of individuals to species.

If you are considering two possible species tree assignments and the posterior distributions of the species.coalescent from the two runs are not overlapping, then you can pick the model that has the higher posterior probability, assuming the species.coalescent is computed up to the same constant for differing numbers of species and species assignments.

Comparing clade ages: If you want to compare the estimated ages for two (non-overlapping) clades, say A and B, probably the best approach is to estimate the posterior probability that the age of clade A is greater than that of B. But the R statistical programming language can compute this with a few lines (assuming a BEAST log file 'post.log', containing columns called 'tmrcaA' and 'tmrcaB'):

```
# read in the log file
post <- read.table("post.log", sep="\t", header=TRUE)
# remove 10 percent burnin
post <- post[round(nrow(post)*0.1+1):nrow(post),]
# calculate posterior probability of tmrcaA > tmrcaB
pp <- mean(post$tmrcaA > post$tmrcaB)
```

Alternatively it is easy enough to load the trace log in which the ages have been reported as MRCA statistics into a spreadsheet. Copy the two columns of interest into a new sheet, and in the third column use '=IF(A1>B1, 1, 0)' and take the average over the last column. This gives you $\Pr(A > B|D)$ where D is the data. Either way, you should remember to remove the burn-in (lines 3–4 in the R script).

As always, it is important to not only report the posterior support for $A > B$ but also the prior support. Without this, the value of $\Pr(A > B|D)$ is harder to interpret in terms of the evidence provided by the data at hand, since the prior can in principle be set up to support any hypothesis over another. If no special tree priors are added and the samples are contemporaneous, we would expect the prior $\Pr(A > B)$ to be 0.5 if A and B are of the same size, indicating 50% of the time one clade should be higher than the other. $\Pr(A > B)$ can be calculated with the same procedure used to calculate $\Pr(A > B|D)$, but this time with the sample from the prior. The BF is then calculated as (Suchard et al. 2001):

Table 10.1 Balanced labelled rooted trees of four taxa are represented by two labelled histories, depending on which cherry (pair of taxa with a common parent) is older

Labelled histories with AB grouping	Labelled histories without AB grouping	
((A,B),C),D)	((A,C),B),D)	((B,D),C),A)
((A,B),D),C)	((A,C),D),B)	((C,D),A),B)
((A,B):1,(C,D):2)	((A,D),B),C)	((C,D),B),A)
((A,B):2 (C,D):1)	((A,D),C),B)	((A,C):1,(B,D):2)
	((B,C),A),D)	((A,C):2 (B,D):1)
	((B,C),D),A)	((A,D):1,(B,C):2)
	((B,D),A),C)	((A,D):2 (B,C):1)

$$BF = \frac{\Pr(D|A)}{\Pr(D|B)} = \frac{\Pr(A > B|D)(1 - \Pr(A > B))}{(1 - \Pr(A > B|D))\Pr(A > B)}, \tag{10.1}$$

which can be interpreted using Table 1.1.

Testing monophyly of a clade: Since the prior probability of a particular clade (given a uniform distribution on labelled histories) is relatively easy to compute, the BF for a particular clade is also easy to compute from the prior and the posterior clade probabilities. For example in BEAST, the prior probability of the (A,B) grouping in a four-taxa tree of {A,B,C,D} is $4/18 = 2/9$, because 4 of the 18 possible labelled histories have the AB grouping, as shown in Table 10.1.

However, if there are calibrations, then the prior of the tree will in general not be uniform over ranked histories. In such cases you can run your model by sampling from the prior and use TreeLogAnalyser or DensiTree to find the clade probability of interest. Note that sampling from the prior may require considerably more samples than sampling from the posterior since the space that needs to be sampled is very large if the priors used are very broad. Fortunately, evaluating each state takes a lot less computational effort than when sampling the full posterior.

Testing sensitivity of parameter estimates to different priors: It is always a good idea to test whether the outcome of the analysis is informed by the data, or just a reflection of the prior. If you suspect that the data have little effect on the parameters of interest, the easiest way to test this is by running the same analysis but sample from the prior. If the marginal prior distribution of the parameter is similar to the marginal posterior distribution, then the data do have much to say about the parameter, and the result will be very sensitive to the choice of prior. To be sure, you can rerun the analysis with a different prior, say, with a different mean. This is a sensitivity analysis.

10.3   Troubleshooting

There are a large number of potential problems that may be encountered when running an MCMC analysis. In this section, we will have a look at some of the more common issues and how to solve them.

Checking for signal in serially sampled data: For serially sampled data, a quick test to see whether there is signal in the data is to construct a neighbour-joining tree and look at the amount of evolution by inspecting the distribution of root-to-tip genetic distances. Path-O-gen[1] (Rambaut 2010) can be useful to show more formally how well divergence correlates with time of sampling (Drummond et al. 2003). If there is no structure, this might indicate that some sequences are misaligned or recombinant sequences, causing larger divergence times than expected. As a consequence estimates for any of the parameters of interest may be different as well.

Fitting an elephant: Our experience is that many practical problems in phylogenetic MCMC with BEAST arise from users immediately trying to fit the most complex model to their data. Though it is tempting to 'fit an elephant' (Steel 2005), a pragmatic approach that starts with a simple model works best in our experience. Only after you have a handle on your data, by having analysed it with a simple model, would we then recommend that you use model selection and model averaging tools to see if a more complex model might be justified (see Sections 10.2 and 9.5).

Non-starting: BEAST might not start because the initial state has a posterior probability of zero, which is an invalid state for the chain to be in. If this happens, BEAST will print out all components that make up the posterior, and the first one that is marked as negative infinity is the component that should be investigated more closely.

There are several reasons why the posterior can be calculated as zero (hence the log posterior as $-\infty$). When using hard bounds in calibrations you need to ensure that the starting tree is compatible with the bounds you have specified. In BEAST 2, randomly generated trees are automatically adjusted to these constraints, as long as the constraints are mutually compatible. The priors can be incompatible if a calibration on a subclade has a lower bound that is higher than the upper bound of a calibration on a superclade. In this situation no tree exists that fits both of these constraints. If for some reason no random starting tree can be found and the calibrations are compatible, you could provide a starting tree in Newick format that is consistent with the calibrations. If such a starting tree is inconsistent, BEAST will start with a log posterior that is $-\infty$ and stop immediately.

Another reason for BEAST to register a zero initial posterior probability is when there are a large number of taxa and the initial tree is very far from the optimum. Then, the tree likelihood will be so small, since the data do not fit the tree, that the tree likelihood can't be calculated due to numerical underflow and the likelihood will return 0 (hence $-\infty$ in log space). To prevent this happening, a better starting tree could be a UPGMA (unweighted pair group method with arithmetic mean) or neighbour-joining tree, or a tree estimated using another program in Newick format that has been adjusted to obey all calibration constraints.

When using an initial clock rate that is many orders of magnitude smaller than the actual clock rate, it can happen that underflow occurs when scaling the branch lengths to units of substitution. A potential reason is that some calibrations are in different units, say millions of years, while the clock rate is in another unit, say years. This is easily fixed by choosing a different starting value for the clock rate.

---

[1] Available from http://tree.bio.ed.ac.uk/software/pathogen.

When BEAST does not start and the coalescent likelihood is reported as $-\infty$, this probably means that a parameter like the growth rate is initially so high that there are numerical issues calculating the coalescent likelihood.

The MCMC chain does not converge: If there are a lot of sequences in the alignment, you need to run a very long chain, perhaps more than 100 million states. Alternatively, you can run a large number of shorter chains, say 20 million states on ten computers. Though a systematic quantitative analysis has yet to be done, in our experience the chain length needs to increase quadratically in the number of sequences. So, if you double the number of sequences then you need to quadruple the length of the chain to get the same ESS. Thus, you could analyse a sub-sample of, say, 20 sequences and use this to estimate how long a chain you need to analyse 40 or 80 or 160 sequences and obtain the same ESS. In general if you have much more than 100 sequences you should expect the analysis to involve a lot of computation and doing multiple runs and combining them is a good idea.

One reason a chain does not converge is when the posterior is multi-modal. Mixing for multi-modal posteriors can take a long time because the operators can have difficulty finding a path through parameter space between the different modes. Since the ESS is chain length divided by autocorrelation time (ACT) and the ACT can be very large for multi-modal traces, the ESS will be very small. Inspecting the traces of parameters in Tracer can help determine if some parameters have two or more modes; however, this is more difficult if the modes are in tree space (Höhna and Drummond 2012; Whidden et al. 2014).

Lack of convergence can also occur if the model is non-identifiable. For example, if both the clock rate and substitution rate are estimated when analysing a single partition, there is no reason to expect the chain to converge, because there are two parameters, but there is only one degree of freedom (i.e. the likelihood only changes as a function of the product of the two parameters). In this case, even if the chain appears to converge, there is no guarantee that the algorithm will produce a sample from the correct target distribution. A model can be made identifiable by choosing fewer parameters to estimate, and in the example above this is achieved by fixing one of the two parameters to, say, 1.0.

In general, more complex models with more parameters converge more slowly because each parameter will be changed less often by operators. Hence, parameter-rich models require more samples to reach the same ESS. There are many more reasons a chain does not converge that are specific to some models. Some of these model-specific problems will be addressed below.

Sampling from the prior does not converge: Sampling from the prior is highly recommended in order to be able to check how the various priors interact. Unfortunately, the space that needs to be sampled for the prior tends to be a lot larger than the posterior space. Therefore, it may take a lot more samples for convergence to be reached when sampling from the prior than when sampling the posterior. It is not uncommon for the difference between the number of samples to be one or two orders of magnitude. Fortunately, evaluating the prior takes a lot less time than evaluating the posterior, so the prior chain should run a lot faster.

It is not possible to sample from the prior if improper priors (see Section 1.5.4) are used. For example, OneOnX priors and uniform priors without bounds can't be used if you want to sample from the prior, or perform path sampling (see Section 1.5.7)

Issues with BSP analysis: When running a BSP analysis, it is not uncommon to see low ESS values for group sizes. The group size is the number of steps in the population size function. These are integer values, and group sizes tend to be highly autocorrelated since they do not change value very often. Unfortunately high ESSs are needed for this parameter to ensure correctness of the analysis.

When there is low variability in the sequences, the estimates for coalescent times will contain large uncertainties and the skyline plot will be hard to reconstruct. So, expect the BSP to fail when there are only a handful of mutations in sequences.

Estimating highly parametric demographic functions from single gene alignments of just a handful of sequences is not likely to be very illuminating. For the single-population coalescent, sequencing more individuals will help, but there is very little return for sequencing more than 20–40 individuals. Longer sequences are better as long as they are completely linked (for example, the mitochondrial genome). Another approach is to sequence multiple independent loci and use a method that can combine the information across multiple loci (like EBSP). Felsenstein (2006) provides an illuminating analysis of the factors contributing to accuracy of coalescent-based estimates. Much of this appears to translate to the context of highly parametric population size estimation (Heled and Drummond 2008).

Demographic reconstruction fails: There are two common reasons for Tracer to fail when performing a demographic reconstruction. Firstly, there may be no 'End;' at the end of the tree log. This can be fixed easily by adding a line to the log with 'End;' in a text editor. Secondly, Tracer fails when the sample frequencies of the trace log and the tree log file differ. The file with the highest frequency can be sub-sampled using LogCombiner.

Relaxed clock gone wrong: Sometimes models using the relaxed clock model take a long time to converge. In general, they tend to take longer than when using a strict molecular clock model with otherwise the same settings. See tips on increasing ESS in Section 10.1 for some techniques to deal with this.

When a relaxed clock model does converge, but the coefficient of variation is much larger than 1, this may be an indication something did not go well. A coefficient of variation of $0.1 < c_v < 1$ is already a lot of variation in rates from branch to branch, especially in an intra-species data set. Having a coefficient of variation of much larger than 1 represents a very large amount of branch rate heterogeneity and if that were really the case it would be very hard to sample from the posterior distribution due to the strong correlations between the divergence times and the highly variable branch rates. It is not much use to try to estimate divergence times under these circumstances. This problem can occur due to a prior on S, the standard deviation of the log rate (when using the log-normal relaxed clock) that is too broad. It may also be due to failure to converge, which can be tested by rerunning the analysis a few times to see whether the same posterior estimates are reached.

If you are looking mostly at intra-species data then the uncorrelated relaxed clock should be used with caution. We recommend using a log-normal uncorrelated relaxed clock with a prior on S centred around $S = 0.1$, and with the bulk of the prior probability below $S = 1$. Alternatively, a random local clock might be more appropriate, since one would expect most of the diversity within a species to be generated by the same underlying evolutionary rate.

Rate estimate gone wrong: If you only have isochronous sequences, that is, all sampled from the same time, and you do not have any extra calibration information on any of the internal nodes nor a narrow prior on clock rates, then there is nothing you can do in BEAST that will create information about rates and dates.

If you are looking at fast-evolving species such as RNA viruses, then a decade of sampling through time is typically more than enough to get good estimates of the rate of evolution. However, if you are looking at a slow-evolving species like humans or mice, this is not a sufficient time interval to allow estimation of mutation rates, unless perhaps if you have sequenced whole genomes. The issue is that with species exhibiting very low evolutionary rates (like mammals), all tip dates become effectively contemporaneous and we end up in the situation with (almost) no calibration information.

RNA viruses like HIV, influenza and hepatitis C have substitution rates of about 10^{-3} (Drummond et al. 2003a; Jenkins et al. 2002; Lemey et al. 2004); however, other viruses can evolve a lot more slowly, some as slowly as 10^{-8} (Duffy et al. 2008). For those slow-evolving viruses there is the same issue as for mammals, and BEAST will not be able to estimate a rate based on samples that have been isolated only decades apart. This can show up in the estimate of the root height having 95% HPDs that are very large. If the data reflect rates that are too slow to estimate, the posterior of the rate estimate should equal the prior, as is indeed the case with BEAST when using simulated data (Firth et al. 2010). However, care is needed because real data can contain different signals causing the rate estimates to be too high due to model misspecification.

Another factor is the number of sampling times available. For example, if you only have two distinct sampling times, far apart compared to the coalescent rate, so that the clades of the old and new sequences are reciprocally monophyletic with respect to each other, then you will not be able to reliably estimate the rate because there is no way of determining where along the branch between the two clusters the root should attach. In this situation, due to the tree topology, there is a 'pulley' at the root which creates large uncertainty in the rate estimates.

One thing you should remember about a Bayesian estimate is that it includes uncertainty. So even if the point estimates of a rate are quite different, they are not significantly different unless the point estimate of one analysis is outside the 95% HPD interval of another analysis with a different prior. In other words, the point estimates under different priors may be different just because the estimates are very uncertain.

Conflicting calibration and clock rate priors: It is possible to have timing information (that is, tip or node dating) while also having an informative prior on the clock rate, or even fixing the clock rate. If the clock rate is constrained to a value considerably different from the rate implied by the (tip or node) calibration information, then convergence could become a problem. In BEAST, genetic distances are modelled by the product of rate and time. If you try to fix both the rate and the time (by having informative tip or node calibration(s)) then there is the potential to get a strong conflict between the genetic distances implied by those data, and those implied by the product of the priors. When a chain has to accommodate such a mismatch, this can manifest itself in poor convergence, unrealistic estimates in parameter values and unexpected tree topologies. It is important to make sure that this situation does not occur inadvertently by fixing the clock rate to 1, when calibration information is available.

Tracer out of memory: Tracer reads the complete log file into memory before processing it, and it can run out of memory in the process. The log files increase proportionally to the number of samples. For very long runs the log intervals should be increased so that the total number of entries that end up in the log is no more than 10 000. So, for a run of 50 million, don't log more often than every 5000 samples. If you have produced a log with too many samples, then use LogCombiner to down-sample the log file before loading it in Tracer.

11 Exploring phylogenetic tree space

11.1 Tree space

In phylogenetic inference the parameters to estimate are the tree topology (or ranked tree) and associated divergence times of the common ancestors. This is not a standard statistical problem because the parameter space is not a simple Euclidean space,[1] i.e. it is not $\mathbb{R}^n$, or any simple convex subspace of $\mathbb{R}^n$. While the space of trees can be considered to be constructed of Euclidean subspaces that do lie in $\mathbb{R}^n$ (one n-dimensional orthant, that is, the restriction of $\mathbb{R}^n$ to non-negative reals, for each tree topology or ranked tree; see Figure 11.1), its overall structure is not Euclidean (see Figure 11.2), and thus new statistical techniques are required for analysis and visualisation.

Formally, a phylogenetic tree space, or just *tree space* for short, is a *metric space* such that the points of the space are in one-to-one correspondence with the set of phylogenetic trees on n taxa (i.e. there is a bijection between the metric space and the set of all trees, where the set of trees has the size of the continuum, since each combination of divergence times on a topology represents a distinct tree). The distance of the metric space induces (via the isomorphism mentioned above) a distance between any pair of trees. Here, as throughout the book, we are specifically considering *time-trees* and *time-tree space*, but we will use the term *tree space* for short.

Despite the fact that statistical inference of phylogenetic trees is a decades-old pursuit, surprisingly little theoretical development of tree metric spaces has taken place and there are many open problems and challenges that derive from the non-Euclidean nature of tree space. A key challenge lies in how one should summarise a set of trees in tree space. Basic concepts in statistics, such as the mean and variance of a sample, are challenging to define for tree space (unless the topology is fixed), and have unusual properties compared to their counterparts in Euclidean spaces. A key result in this area was the description of the geometry of tree space for phylogenetic trees with unconstrained branch lengths (i.e. not time-trees), called BHV space (Billera et al. 2001). Recently there have been a number of exciting developments regarding BHV space, including the description of a polynomial-time algorithm to compute distances between trees in BHV space (Owen and Provan 2011), opening up new approaches to producing Bayesian point estimates of phylogenetic reconstructions (Benner et al. 2014). However, no such analogous results are available for time-tree space.

---

[1] Although those familiar with differential geometry or philosophy might disagree that Euclidean space is simple!

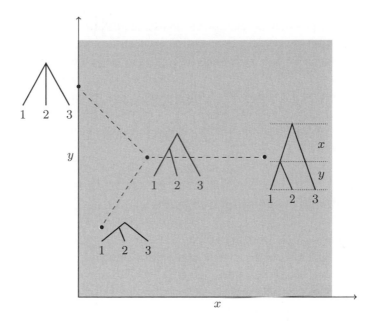

Figure 11.1 A Euclidean two-dimensional space representing the space of all possible time-trees for the topology ((1,2),3). There are two parameters, x and y, one for each of the two inter-coalescent intervals, the sum of which is the age of the root ($t_{root} = x + y$). Three trees are displayed, along with their arithmetic mean tree, also called the *centroid*. The dashed lines show the path connecting each of the three trees to the mean tree by the shortest distance (i.e. their deviations from the mean).

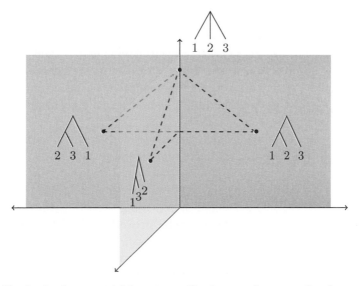

Figure 11.2 $\mathbb{T}_3$, the simplest non-trivial tree space (for time-trees), representing the space of time-trees for $n = 3$ taxa sampled contemporaneously. Each of the three non-degenerate tree topologies is represented by a two-dimensional Euclidean space (as illustrated in Figure 11.1) and these subspaces meet at a single shared edge representing the star tree, which is a one-dimensional subspace and thus has a single parameter (the age of the root). The dashed lines shows the paths of shortest distance between the four displayed trees.

It is precisely because of the non-standard nature of tree space, and the consequent limits to its statistical characterisation, that there are specialist programs for Bayesian phylogenetic inference. Otherwise, popular general tools for Bayesian statistical inference such as Stan (Hoffman and Gelman 2014; Stan Development Team 2014) or BUGS (Lunn et al. 2000, 2009) could be used. This statement implies that Stan or BUGS *could* be productively employed for Bayesian phylogenetic inference problems that are conditioned on a fixed topology, or ranked history, depending on the tree space employed.

11.2   Methods of exploring tree space

A Bayesian phylogenetic MCMC analysis produces one or more tree log files containing a sample of trees from the posterior distribution over tree space. The MCMC algorithm produces a chain of states, and these are autocorrelated so that in general two adjacent states in the chain are not independent draws from the posterior distribution. Thus a critical element of evaluating the resulting chain of sampled trees from a Bayesian phylogenetic MCMC analysis is determining whether the chain is long enough to be representative of the full posterior distribution over tree space (Nylander et al. 2008). In general it is not trivial to quantify MCMC exploration of phylogenetic tree space (Whidden et al. 2014) and careful investigation can reveal unexpected properties. Research into efficient sampling of phylogenetic tree space, including improving existing MCMC algorithms, is an active field of enquiry (Höhna and Drummond 2012; Lakner et al. 2008; Whidden et al. 2014). In this chapter we will primarily consider what one does after a representative posterior sample of trees has been obtained. In practice this stage is reached by running a long enough chain (or multiple chains) so that standard diagnostic tests are passed, and then taking a regular subsampling of the full chain(s) as your resulting posterior sample (i.e. the contents of the tree log file(s)).

The main methods for dealing with such a posterior sample of trees are:

1. enumerating the most commonly (co-)occurring clades/topologies,
2. summarising the sample with a single point estimate (i.e. a summary tree),
3. performing multi-dimensional scaling using a tree metric,
4. visualising the sample of trees in a *tree space graph*,
5. visualising the sample of trees using DensiTree (Bouckaert 2010) and
6. constructing a phylogenetic network that contains the common clades/trees (not described).

In this chapter we briefly review the first five of these methods and compare them. The methods are judged on their ability to clarify properties of the posterior sample, and in particular whether they can highlight areas of certainty and uncertainty in both divergence times and tree topology. We will pay special attention to summary trees and DensiTrees. Visualising phylogenies and phylogenetic tree space are active areas of research (Graham and Kennedy 2010; Procter et al. 2010; Whidden et al. 2014).

As an illustrative example, we use sequences from *Anolis*, a genus of lizards belonging to the family Polychrotidae. The alignment can be found in the BEAST directory with NEXUS examples and has been published in (Jackman et al. 1999).[2] There are 55 taxa and the nucleotide alignment has 1456 sites (Matrix M3924 in TreeBase). To produce the tree set, the first 1035 sites were partitioned into codon positions 1, 2 and 3. The remaining sites code for a tRNA gene. The substitution model used for each of the four partitions was GTR with four gamma-distributed site rate categories. Furthermore, a strict molecular clock and coalescent tree prior with exponential population growth were used.

11.3   Tree set analysis methods

In this section, we consider some tree set analysis methods. Some work equally well for both rooted and unrooted trees, while others are designed only for rooted trees.

11.3.1   Credible sets of phylogenetic trees

For data sets that are well resolved, one might consider reporting not a single summary tree, but instead a set of topologies that together account for a large fraction of the posterior probability. This is known as a credible set of tree topologies and 95% is often used as the threshold fraction. The 95% credible set of topologies is the smallest set of topologies whose total posterior probability is ≥ 0.95.

For example, for the primate analysis from Chapter 6, the 95% credible set contains a single tree topology. For *Anolis*, we ran four independent MCMC analyses to demonstrate the variability of the credible set estimates. The most probable topology had a estimated posterior probability of 0.0139 ± 0.0006 (i.e. 1.39% of the sampled trees across the four runs). The estimate of its posterior probability varied across runs (0.0126, 0.0152, 0.0147, 0.0132), but the same topology always had the highest posterior probability. This is despite the fact that the 95% credible set had around 8300 ± 110 unique topologies (8263, 8428, 8042, 8573). Although in the case of the *Anolis* data the 95% credible set is not useful for visual inspection, it can still be a useful tool for assessing if an *a priori* phylogenetic hypothesis is supported by the data.

11.3.2   Clade sets

A non-graphical method of tree set analysis is achieved by inspecting the posterior support for clades (Mau et al. 1999), for instance by listing the most frequently occurring clades. For the tree set consisting of the three trees in Figure 11.3 the clade that groups taxa {A,B,C} occurs three times, but the clades {A,B}, {A,C} and {B,C} occur only once each. In this approach, a list of clades ordered by frequency are generated and the most frequently occurring clades interpreted as the ones best supported by the data. The proportion of times a subset of taxa appears as a monophyletic group

---

[2] Available from http://purl.org/phylo/treebase/phylows/study/TB2:S603?format=html.

Figure 11.3 Small example of a tree set.

in the posterior sample is a Monte Carlo estimate of the clade's posterior probability. These are sometimes termed *posterior clade probabilities* or *posterior clade support*. The problem with this approach is that, $C(n)$, the number of potential non-trivial clades (i.e. strict subsets of taxa of size ≥ 2), grows exponentially with the number of taxa, n: $C(n) = 2^n - n - 2$.

Although the posterior support can often be localised in a small fraction of tree space, if everything else is equal we may well assume that the clades with significant posterior support might also grow roughly exponentially with the number of taxa analysed. When there are many closely related taxa in a data set, the number of clades appearing in 5% or more of the posterior sample can easily be in the hundreds or thousands. Obviously, such clade sets would be hard to interpret without good visualisation. Another problem is that credible sets only provide information about the uncertainty in the tree topology, and do not inform about the uncertainty in divergence times, unless each clade is additionally annotated with information about the marginal posterior distribution of the age of the clade's most recent common ancestor (e.g. 95% HPD intervals).

As an illustration, a typical posterior sample from the *Anolis* data set of 18 000 trees (20 000 trees minus 10% burn-in) would comprise 9200 ± 110 unique topologies (9163, 8942, 9328, 9473). More comfortably, the 5% credible set of topologies contains only 58 different clades (all four runs agreed). The 50% credible set includes only about 107 ± 2.7 clades (103, 106, 105, 115) and the 95% credible set includes only 156 ± 4.1 clades (158, 152, 148, 167). These are all quite small numbers compared to the approximately 36 million billion possible clades for a 55-taxa tree. So, while there are a large number of tree topologies, a relatively small number of clades dominate the tree distribution. In a typical run of the *Anolis* data there are 41 clades that each occur in over 90% of the trees. In fact (*A. angusticeps, A. paternus*) is one of 13 two-taxa clades that have a posterior probability of 1.0. So the topological uncertainty is limited to a small number of divergences, and a small number of alternatives at those divergences. But because there is uncertainty present in a number of different parts of the phylogeny, there is a combinatorial explosion of full resolutions of the tree, leading to the large number of unique trees in the posterior sample and credible set.

While it is easy to provide a list of topologies and clades and the number of times they appear in the posterior sample (for example, with the TreeLogAnalyser tool in BEAST), such data are difficult to interpret, and it does not give much intuition about the overall structure of the posterior distribution. However, it can be useful for testing specific *a priori* phylogenetic hypotheses such as whether a clade is monophyletic (see Section 10.2).

11.3.3 Conditional clades

An improvement over simple estimation of the posterior probability of individual tree topologies has been recently suggested (Höhna and Drummond 2012; Larget 2013). It is good when the posterior distribution is more diffuse, or when one is interested in the probability of improbable trees. Instead of using the simple Monte Carlo estimate (i.e. the proportion of times the tree is sampled in the MCMC chain), one can use the product of the probabilities of the conditional clades that the tree is composed of. These *conditional clade probabilities* are calculated by simple Monte Carlo estimates (i.e. their proportion in the MCMC chain). However, because the number of conditional clades is much smaller than the number of trees for large n, the estimates of conditional clade probabilities are expected to be more accurate than direct Monte Carlo estimates of individual tree topologies. This approach also allows the estimation of the posterior probability of trees that were not sampled, as long as the conditional clades that the tree is composed of are present in the posterior sample. Although this approach appears to have some empirical support in the two papers that introduce it, other studies have cast some doubt as to whether this method provides an improvement over alternatives (Heled and Bouckaert 2013).

11.3.4 Multi-dimensional scaling

Multi-dimensional scaling (MDS) is a technique for visualising items in a high-dimensional data set when there is a *distance metric* defined between these items (Amenta and Klingner 2002; Hillis et al. 2005). In our situation, these items are individual trees in the sample. There are many possible distance metrics between trees. Some take into account only the topology (e.g. the minimum number of tree edit operations required to convert one topology into another), while others take into account the divergence times as well (e.g. the average difference in pairwise path length of the trees). See (Heled and Bouckaert 2013) for more on time-tree distance metrics. Figure 11.4 shows the visualisation of the *Anolis* data, drawn with the method from (Hillis et al. 2005). Each tree represents a point in the image and the farther two points are apart, the greater the dissimilarity of the associated trees. Though this works fairly well when the posterior estimate of the tree is well resolved, in the case of the *Anolis* data, there are almost as many unique topologies as sampled trees. As a result the image looks like white noise and the different groups are not visible in the picture. Furthermore, it is not possible to judge uncertainty in node height from the image, nor is the topology shown.

11.4 Summary trees

There are many ways to construct a summary tree, also known as a consensus tree, from a set of trees. See the excellent review of (Bryant 2003) for a few dozen methods. Most of these methods create a representation of the tree set that are not necessarily good estimators of the phylogeny (Barrett et al. 1991). A few of the more popular options are the following.

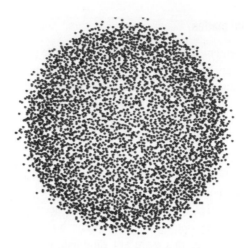

Figure 11.4 Multi-dimensional scaling result for the *Anolis* tree set.

The *majority rule consensus tree* is a tree constructed so that it contains all of the clades that occur in at least 50% of the trees in the posterior distribution. In other words it contains only the clades that have a posterior probability of $\geq 50\%$. The extended majority tree is a fully resolved consensus tree where the remaining clades are selected in order of decreasing posterior probability, under the constraint that each newly selected clade be compatible with all clades already selected. It should be noted that it is quite possible for the majority consensus tree to be a tree topology that has never been sampled and in certain situations it might be a tree topology with relatively low probability, although it will have many features that have quite high probability. Holder et al. (2008) argue that the majority rule consensus tree is the optimal tree for answering the question 'What tree should I publish for this group of taxa, given my data?' assuming a linear cost in the number of incorrect and missing clades with higher cost associated to missing clades.

The *maximum clade credibility* (MCC) tree produced by the original version of TreeAnnotator is the tree in the posterior sample that had the maximum *sum* of posterior clade probabilities. However, we will restrict our use of the term MCC to the more natural current default of TreeAnnotator, which is the tree with the maximum *product* of posterior clade probabilities. The MCC tree is always a tree in the tree set, and is often shown in publications that use BEAST for phylogenetic reconstruction. Recent empirical experiments show that MCC trees perform well on a range of criteria (Heled and Bouckaert 2013).

The term *maximum a posteriori tree* or *MAP tree* has a number of interpretations. It has sometimes been used to describe the tree associated with the sampled state in the MCMC chain that has the highest posterior probability density (Rannala and Yang 1996). This is problematic, because the sampled state with the highest posterior probability density may just happen to have extremely good branch lengths on an otherwise fairly average tree topology. A better definition of the MAP tree topology is the tree topology that has the greatest posterior probability, averaged over all branch lengths and substitution parameter values. For data sets that are very well resolved, or have a small number of taxa, this is easily calculated via Monte Carlo, by just determining which tree topology has been sampled the most often in the chain. However, for large data sets it is quite possible that every sampled tree has a unique topology. In this case, conditional clade probabilities can be used to estimate the MAP topology (Höhna and Drummond 2012; Larget 2013).

A natural candidate for a point estimate is the tree with the maximum product of the posterior clade probabilities, the so-called *maximum credibility tree*. To the extent that the posterior probabilities of different clades are additive, this definition is an estimate of the total probability of the given tree topology, that is, it provides a way of estimating the maximum a posteriori tree (MAP) topology. Höhna and Drummond (2012) and Larget (2013) introduce a method to estimate posterior probabilities of a tree based on conditional clade probabilities.

To define a *median tree*, one must first define a metric on tree space. This turns out to be quite a difficult task to perform, but there are a number of candidate metrics described in the literature. Visualise the trees in the posterior sample as a cluster of points in a high-dimensional space, then the median tree is the tree in the middle of the cluster – the median tree has the shortest average distance to the other trees in the posterior distribution. With a metric defined (say the Robinson–Foulds distance; Robinson and Foulds 1981), a candidate for the median tree would be the tree in the posterior sample that has the minimum mean distance to the other trees in the sample.

Once a tree topology or topologies are found that best summarises a Bayesian phylogenetic analysis, the next question is what divergence times (node heights) to report. One obvious solution is to report the mean (or median) divergence time for each of the clades in the summary tree. This is especially suitable for the majority consensus tree and the maximum credibility tree, however defined. For the median tree, it should be noted that some metrics, that is, those that take account of branch lengths and topology, allow for a single tree in the sample to be chosen that has the median topology and branch lengths. Likewise, if the MAP sampled state is the chosen tree topology, then the associated branch lengths of the chosen state can be reported.

Instead of constructing a single summary tree, a small set of summary trees can be used to represent the tree set (Stockham et al. 2002).

11.4.1   Tools for summary trees

TreeAnnotator is a tool provided with BEAST to generate summary trees. It can create a tree with the topology of the tree set with highest product (or sum) of clade probabilities or annotate a user-provided tree. Sometimes, the summary tree has negative branch lengths. This happens when the clade used to estimate a node height contains just a few trees. Then, height estimates have large variability and nodes can be assigned a height that is higher than its parent, resulting in negative branch lengths. Note that this is not a software bug but, more than anything, this is an indication that there is large uncertainty in the tree topology and node height estimates.

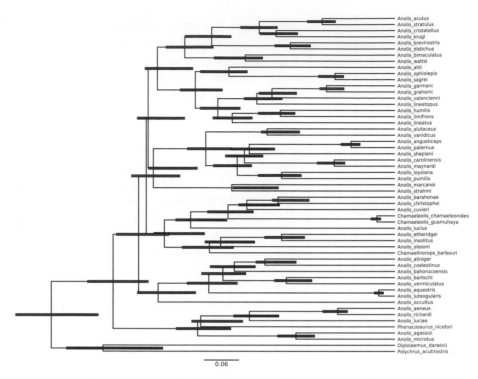

Figure 11.5 Single consensus tree of the *Anolis* tree set, with bars representing uncertainty in node height.

FigTree (available from http://tree.bio.ed.ac.uk/software/figtree/) is a very useful tool for visualising summary trees and producing high-quality output ready for publication. FigTree can export PDF which can be converted to SVG and imported into drawing programs such as CorelDraw to add annotations manually. Another useful function of FigTree is that it allows converting files to Newick format; load a tree, choose export and specify the tree format.

TreeAnnotator only annotates those clades that occur with a certain support in the tree set. This assures that 95% HPD intervals will be reasonably accurate, since 95% HPD intervals will have very high variance when estimated from the information of just a few tree examples. Consequently, some 95% HPD intervals may not show up when visualising the tree in FigTree.

Figure 11.5 shows a consensus tree produced with TreeAnnotator and visualised in FigTree for the *Anolis* data. The tree is annotated with bars representing uncertainty in node heights. This shows it is possible to make uncertainty in node heights visible. However, it is not quite clear whether the source of this uncertainty is due to uncertainty in node heights or because there is no certainty in the underlying topology. For example, the consensus tree of the three-trees example of Figure 11.3 would be any of the trees in the set, but with high uncertainty on the node heights for the common ancestors of clade (A,B,C). Suppose the first tree in Figure 11.3 is chosen as topology, then the 95%

HPD interval for the height with many methods will be of size zero, since the interval will be based on all (A,B) clades in the tree set, and this clade only occurs once. This illustrates the danger of basing estimates on only the clades occurring in the summary tree, and ignoring other information in the tree set.

Uncertainty in topology does not show up in the consensus tree. An alternative is to label the tree nodes with the support in the tree set. In this case, every node would have a number attached with the number of trees that contain the clade associated with that node. A low number could indicate low consensus in the topology. For the *Anolis* tree, most branches have support of over 90%, indicating most of the tree topology is strongly supported by the data. However, for those branches with lower support, it is not clear what the alternative topologies are. In summary, distinguishing between uncertainty in topology and uncertainty in branch lengths requires careful examination of the tree and its annotations.

Algorithms for generating summary trees that use not only those clades in the posterior that are found in the summary tree but that use all clades were developed recently (Heled and Bouckaert 2013). These algorithms do not suffer from negative branch lengths. An implementation is available in biopy (available from https://code.google.com/p/biopy), which is integrated with DensiTree (see next section). One set of algorithms tries to match clade heights as closely as possible, resulting in large trees with long branches. Another set of algorithms tries to match branch lengths. These methods have a tendency to collapse branches that have little clade support, but tend to have higher likelihood on the original data used in simulation experiments.

The *common ancestor heights* algorithm determines the height of nodes in a summary tree by using the average height of the summary tree clades in the trees of the tree set. This method tends to do well in estimating divergence times and is fast to compute (Heled and Bouckaert 2013). Since the height of a node of a clade is always at least as high as any of its subclades, this guarantees that all branch lengths are non-negative. It is the default setting in TreeAnnotator in BEAST 2.2.

11.5   DensiTree

A DensiTree (Bouckaert 2010) is an image of a tree set where every tree in the set is drawn transparently on top of each other. The result is that areas where there is large consensus on the topology of the tree show up as distinct, fat lines while areas where there is no consensus show as blurs. The advantage of a DensiTree is that it is very clear where the uncertainty in the tree set occurs, and no special skills are required to interpret annotations on the tree.

Figure 11.6 shows the DensiTree for the *Anolis* data, together with a consensus tree. The image clearly shows there is large consensus of the topology of the trees close to the leaves of the tree. Also, the outgroup consisting of the clade with *Diplolaemus darwinii* and *Polychrus acutirostris* at the bottom of the image is clearly separated from the other taxa. Again, this might not quite be what is expected from the large number of different topologies in the *Anolis* tree set noted in Section 11.3.2. Where

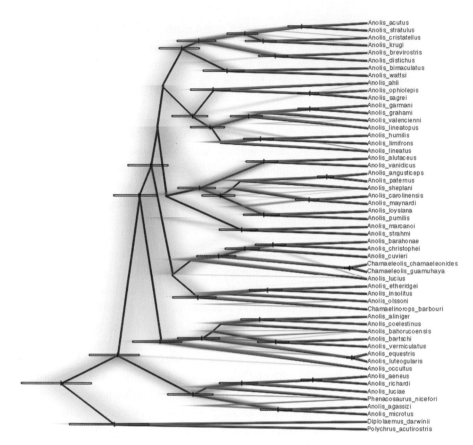

Anolis_acutus
Anolis_stratulus
Anolis_cristatellus
Anolis_krugi
Anolis_brevirostris
Anolis_distichus
Anolis_bimaculatus
Anolis_wattsi
Anolis_ahli
Anolis_ophiolepis
Anolis_sagrei
Anolis_garmani
Anolis_grahami
Anolis_valencienni
Anolis_lineatopus
Anolis_humilis
Anolis_limifrons
Anolis_lineatus
Anolis_alutaceus
Anolis_vanidicus
Anolis_angusticeps
Anolis_paternus
Anolis_sheplani
Anolis_carolinensis
Anolis_maynardi
Anolis_loysiana
Anolis_pumilis
Anolis_marcanoi
Anolis_strahmi
Anolis_barahonae
Anolis_christophei
Anolis_cuvieri
Chamaeleolis_chamaeleonides
Chamaeleolis_guamuhaya
Anolis_lucius
Anolis_etheridgei
Anolis_insolitus
Anolis_olssoni
Chamaelinorops_barbouri
Anolis_aliniger
Anolis_coelestinus
Anolis_bahorucoensis
Anolis_bartschi
Anolis_vermiculatus
Anolis_equestris
Anolis_luteogularis
Anolis_occultus
Anolis_aeneus
Anolis_richardi
Anolis_luciae
Phenacosaurus_nicefori
Anolis_agassizi
Anolis_microtus
Diplolaemus_darwinii
Polychrus_acutirostris

Figure 11.6 DensiTree of the *Anolis* tree set. Bars indicate 95% HPD of the height of clades. Only clades with more than 90% support have their bar drawn.

the topology gets less certain is in the middle, as indicated by the crossing line in the DensiTree.

Consider the clade just above the outgroup consisting of *A. aeneus*, *A. richardi*, *A. luciae*, *Phenacosaurus nicefori*, *A. agassizi* and *A. microtus*. The first three form a solid clade, say clade X, and the last two, say clade Y, as well. However, it is not clear where *P. nicefori* fits in. There are three options; *P. nicefori* split off before the other two clades, clade X split off before *P. nicefori* and clade Y, or clade Y split off before *P. nicefori* and clade X. There is support in the data for all three scenarios, though the last scenario has most support with 50%, judging from the 50% clade probability consisting of clade X and *P. nicefori*. Further, there is 35% support for the first scenario, leaving 15% support for the second scenario. So, where the summary tree shows the most likely scenario, there are two other scenarios in the 95% credible set of scenarios and these are visualised in the image.

However, when a lot of taxa are closely related there can be a lot of uncertainty inside clades. This is shown in Figure 11.7, which shows the tree set for dengue-4 virus. Many samples were used in this analysis that only differ by a few mutations from other

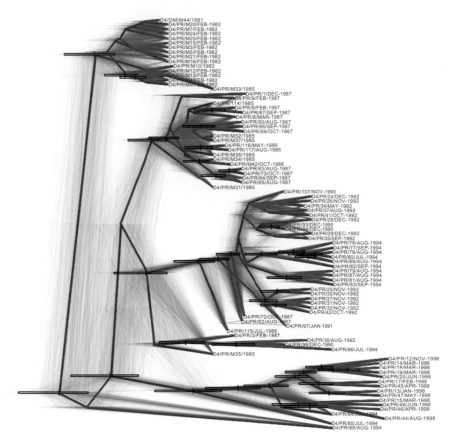

Figure 11.7 DensiTree of the dengue-4 tree set. Bars as in Figure 11.6.

sequences in the alignment. Consequently, there is large uncertainty in the topology of the tree close to the leaves. This is reflected in the DensiTree by the cloud forming around closely related taxa. On the other hand, the remainder of the tree that is not close to the leaves is very well supported by the data. Bars are drawn for those clades that have over 90% support, and all clades higher up in the summary tree are marked with a bar, so they have over 90% support.

A DensiTree highlights the uncertainty in node heights, both by the distribution of the lines forming the trees and bars on the nodes indicating 95% HPD node heights. Furthermore, a DensiTree highlights the uncertainty in topology and often a few clear alternatives are highlighted. It gives some intuition of what a tree distribution produced by a BEAST analysis actually is. So, a DensiTree shows structure and uncertainty in topology and node heights. However, some trees contain just not enough structure and other methods, in particular summary trees, can help visualise the little structure that is available.

Some applications of DensiTree can be found in (Chaves and Smith 2011; Dincă et al. 2011; McCormack et al. 2011).

Part III

Programming

12 Getting started with BEAST 2

BEAST is software for performing a wide range of phylogenetic analyses. The vision we have for BEAST is that it provides tools for computational science that are

1. *easy to use*, that is, well documented, having intuitive user interfaces with shallow learning curve.
2. *open access*, that is, open source, open XML format, facilitating reproducibility of results, and running on many platforms.
3. *easy to extend*, by having extensibility in their design.

We limit the scope of BEAST to efficient Bayesian computation for sequence data analysis involving tree models. Making BEAST easy to use is one of the things that motivated writing this book. The code is set up to encourage documentation that is used in user interfaces like BEAUti. By dividing the code base into a core set of classes that can be extended by packages (Chapter 15), we hope that it will be easier for new developers to learn how to write new functionality and perform new science. Further help and documentation is available via the BEAST 2 wiki.[1]

We want BEAST to be open access (Vision 2) and therefore it is written in Java, open source and licensed under the Lesser GNU Public License.[2] BEAST 2 typically runs as a standalone application, by double-clicking its icon (in modern operating systems) or starting from the command line with `java -jar beast.jar`. A BEAST 2 XML file should be specified as the command line argument. XML files are used to store models and data in a single place. The XML format is an open format described in Chapter 13.

Since we want the system to be extensible (Vision 3), everything in the system implements `BEASTInterface`. The `BEASTObject` class provides a basic implementation and many classes in BEAST derive from `BEASTObject`. We will say that an object is a BEAST-object if it implements `BEASTInterface` and an object is a `BEASTObject` if it derives from `BEASTObject`. Every BEAST-object can specify inputs to connect with other BEAST-objects, which allows for flexible model building. `Input` objects contain information on type of input and how they are stored in BEAST XML files. To extend the code, you write Java classes that implement `BEASTInterface` by deriving from the `BEASTObject` class, or deriving from any

---

[1] http://beast2.org/wiki
[2] The BEAST 2 source code can be downloaded from http://beast2.org.

of the more specialised classes that subclass `BEASTObject`. But before we get into the gory details of writing BEAST-objects, let us first have a guided tour of BEAST 2.

12.1   A quick tour of BEAST 2

Figure 12.1 shows (part of) a model, representing a nucleotide sequence analysis using the Jukes–Cantor substitution model. The 'rockets' represent BEAST-objects, and their 'thrusters' the inputs. Models can be built up by connecting BEAST-objects through these inputs with other BEAST-objects. For example, in Figure 12.1 the `SiteModel` BEAST-object has a `JC69` substitution model BEAST-object as input, and `Tree`, `SiteModel` and `Alignment` are inputs to the `TreeLikelihood` BEAST-object. The `TreeLikelihood` calculates the likelihood of the alignment for a given tree. To do this, the `TreeLikelihood` also needs at least a `SiteModel` as input, and potentially also a `BranchRateModel` (not necessary in this example and a strict clock is assumed by default). The `SiteModel` specifies everything related to the transition probabilities for a site from one node to another in the `Tree`, such as the number of gamma categories, proportion of invariant sites and substitution model. In Figure 12.1, the Jukes–Cantor substitution model is used. In this section, we extend this with the HKY substitution model and show how this model interacts with the operators, state, loggers and other bits and pieces in the model.

To define the HKY substitution model, first we need to find out what its inputs should be. The kappa parameter of the HKY model represents a variable that can be estimated. BEAST-objects in the calculation model (that is, the part of the model that performs the posterior calculation) are divided into `StateNodes` and `CalculationNodes`. `StateNodes` are classes an operator can change, while `CalculationNodes` are classes that change their internal state based on inputs. The HKY model is a `CalculationNode`, since it internally stores an eigenvalue matrix that is calculated based on kappa. Kappa can be changed by an operator and does not calculate anything itself, so the kappa parameter is a `StateNode`.

The other bit of information required for the HKY model is the character frequencies. These can be calculated from the alignment or estimated using a parameter, so

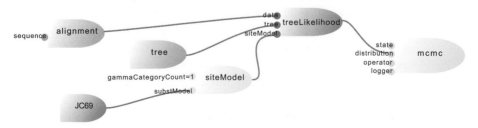

Figure 12.1 Example of a model specifying the Jukes–Cantor substitution model (JC69). It shows BEAST-objects represented by rocket shapes connected to other BEAST-objects through inputs (the thrusters of the rocket).

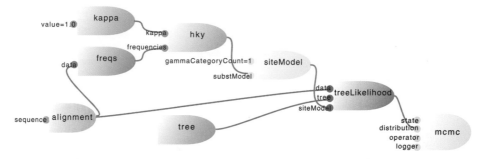

Figure 12.2 Example of a model specifying a HKY substitution model.

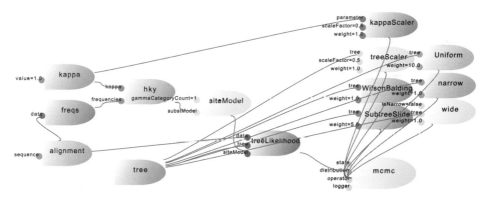

Figure 12.3 Adding operators.

`Frequencies` is a `CalculationNode`. Compare Figure 12.2 with Figure 12.1 to see how the HKY model differs from the JC model. Section 12.2 has implementation details for BEAST-object classes like HKY.

In an MCMC framework, operators propose a move in a state space, and these are then accepted or rejected based on how good the moves are. Figure 12.3 shows the HKY model extended with seven operators: six for changing the tree and one for changing the kappa parameter.

The operators work on the tree and the kappa parameter. Any `StateNode` that an operator can work on must be part of the `State` (Figure 12.4). Apart from the `State` being a collection of `StateNodes`, the `State` performs introspection on the model and controls the order in which BEAST-objects are notified of changes and which of them should store or restore their internal state. For example, if an operator changes the Tree, the HKY model does not need to be bothered with updating its internal state or storing that internal state since it never needs to be restored based on the tree change alone.

For the MCMC analysis to be any use, we need to log the results. Loggers take care of this task. Loggers can log anything that is `Loggable`, such as parameters and trees, but it is easy enough to write a custom logger and add it to the list of inputs of a Logger (Figure 12.5). Typically, one logger logs to standard output, one to a log file

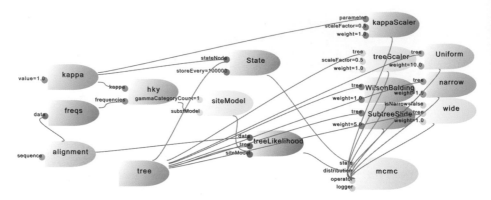

Figure 12.4 Adding the state.

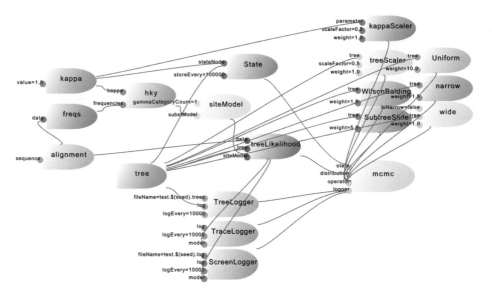

Figure 12.5 Adding the loggers.

with parameter values (a tab delimited file that can be analysed with Tracer) and one log file with trees in Newick format.

Finally, the `Alignment` consists of a list of `Sequences`. Each sequence object contains the actual sequence and taxon information. This completes the model, shown in Figure 12.6 and this model can be executed by BEAST 2.

However, this does not represent a proper Bayesian analysis, since no prior is defined. We need to define one prior for each of the items that form the `State`, in this case a tree and the kappa parameter of the HKY model. A posterior is a `Distribution` that is the product of prior and likelihood, which themselves are a `Distribution`. For such distributions there is the `CompoundDistribution`, and Figure 12.7 shows the complete model with posterior, prior and likelihood as `CompoundDistribution` BEAST-objects. The prior consists of a log-normal prior on the kappa parameter and a Yule prior on the tree. Since the Yule prior has a birth rate that can be estimated, the

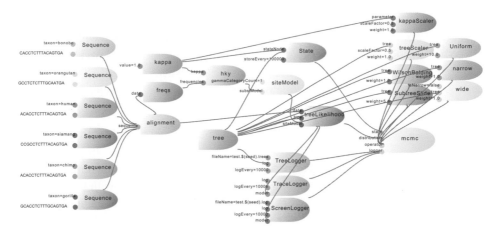

Figure 12.6 Adding the sequences. This forms a complete description of the model, which can be executed in BEAST 2.

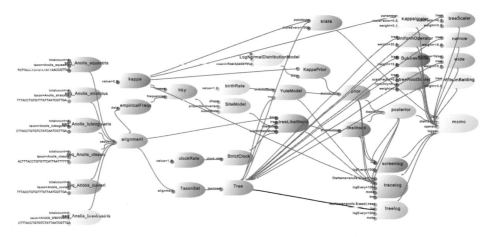

Figure 12.7 Adding a posterior, prior and likelihood and appropriate priors on kappa, the tree and birth rate of the Yule prior. This model forms a proper Bayesian analysis that can be executed in BEAST 2.

birth rate parameter requires a prior as well (and is part of the `State`, not shown in the figure). The prior on the birth rate is a uniform prior in Figure 12.7.

12.2 BEAST core: BEAST-objects and inputs

One way of looking at BEAST is that it is a library consisting of two parts: an MCMC library, which lives in the `beast.core` package, and an evolution library in the `beast.evolution` package. Beast, BEAUti, SequenceGenerator and a handful of other tools are applications built on top of these libraries, and the application-specific code is in the `beast.app` package. Since all computational science heavily relies on mathematics, a `beast.math` package containing a variety of mathematical functionality could not be avoided. There is also a package with assorted utilities, such as file parsers, formatters, random number generation and package management in the `beast.util` package. Together, these packages form BEAST.

The BEAST 2 philosophy is 'everything is a BEAST-object'. A BEAST-object is an object that provides the following:

- *graph structure*: BEAST-objects are connected with other BEAST-objects through 'inputs', which represent links between BEAST-objects. This way BEAST-objects form a directed acyclic graph as shown in Figures 12.1–12.7.
- *validation*: inputs have validation rules associated with them, such as whether an input is optional or required, which triggers some automatic sanity checks for a model. The graph structure makes it possible to validate aspects of the model systematically. For example, it is easy to check that every operator should operate on a `StateNode` that is in the `State`.
- *documentation*: through `Description` and `Citation` annotations, and descriptions and names on inputs.
- *XML parsing*: BEAST-objects together with their input define an XML file format so models can be written in readable format.

The task of a BEAST-object writer is to create classes, specify inputs and provide extra validation that is not already provided by inputs. Chapter 14 goes into the details of the coding involved in writing BEAST-objects.

12.3   MCMC library

Bayesian computation is most often accomplished using the MCMC algorithm. Box 12.1 shows the basic structure of the MCMC algorithm. A glance at this bit of pseudo-code reveals that the least that is required are the following components:

- a data/alignment object that contains sequence data;
- a state object to represent the current and proposed state. The state consists of at least one tree (see scope at the start of this chapter) and parameters, which can be integer, real or boolean valued;
- probability distribution object to calculate the posterior;
- operator objects to work on the state and propose new states;
- log objects, since we are interested in the results, which have to be recorded somewhere;
- an MCMC object to control the flow of the computation.

So, that leaves us with at least an `Alignment`, `State`, `Parameter`, `Tree`, `Distribution`, `Operator`, `Logger` and `MCMC` object. Since trees consist of nodes, a `Node` object is desirable as well. Further, to ensure type safety we distinguish three types of parameter, `RealParameter`, `IntegerParameter` and `BooleanParameter`.

Box 12.1 Basic structure of MCMC.

```
 1 Read data
 2 Initialize state
 3 while (not tired) {
 4      Propose new state
 5      calculateLogPosterior();
 6      if (new state is acceptable)
 7          // do something
 8      else
 9          // do something else
10      Log state
11 }
```

12.3.1   MCMC/runable

A good understanding of the implementation of the MCMC algorithm in BEAST is essential in writing efficient BEAST-objects. In this section we will go through its details. The main loop of the MCMC algorithm executes the following steps after the state is initialised:

```
 1 Propose new state
 2 logP = calculateLogPosterior();
 3 if (new state is acceptable)
 4      // do something
 5 else
 6      // do something else
```

In principle, it explores the space (represented by the state) by randomly proposing new states in the state space. The quality of the state is determined by the posterior probability, which can be quite small, hence we calculate the logarithm of the posterior to prevent numeric underflow. Depending on the acceptance criterion, the new state is accepted or rejected. Where the MCMC algorithm differs from simulated annealing, random hill climbing and other optimisation algorithms is in the acceptance criterion and underlying theory, which allows us to interpret the points visited in state space as a sample from the posterior distribution. Since MCMC algorithms may require a long time to converge, it is important to implement the evaluation of the proposed new state as well as accepting and rejecting a new state as efficiently as possible. First, we will have a look at how this affects the state, as shown in the following listing:

```
 1 Store state
 2 Propose new state
 3 logP = calculateLogPosterior();
 4 if (new state is acceptable)
 5      accept state
 6 else
 7      restore state
```

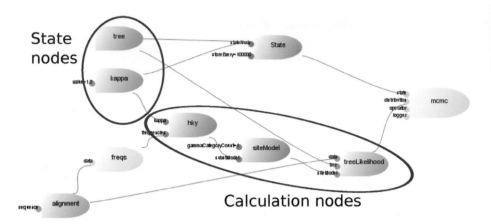

Figure 12.8 The difference between `StateNode` and `CalculationNode`. `StateNode`s are part of the `State` and can only be changed by operators. `CalculationNode`s change when one of it's input is a `StateNode` that changed or a `CalculationNode` that changed.

At the start of the loop in line 1 the state is stored, which makes it easy to restore the state later, if required. When a new state is proposed (line 2), one or more of the `StateNode`s in the state will be given a new value; for example, a parameter may have its values scaled or a tree may have its topology changed. The `State` keeps track of which of the `StateNode`s are changed. All `StateNode`s that changed have a flag marking that they are 'dirty', while all other `StateNode`s are marked as 'clean'. If the state turns out to be acceptable, the state is notified (line 5) and all `StateNode`s that were marked dirty before are now marked clean again. If the state is not acceptable, the state should be restored to the old state that was stored in line 1.

The `State` is aware of the network of BEAST-objects and can calculate which of the `CalculationNode`s may be impacted by a change of a `StateNode`. Figure 12.8 shows a `State` consisting of a kappa parameter and a tree. When the kappa parameter changes, this has an effect on the HKY substitution model, which causes a change in the site model, which requires the `TreeLikelihood` to be updated. However, if the tree changes, but the kappa parameter remains the same, there is no need to update the HKY model or the site model. In fact, the `TreeLikelihood` is set up to detect which part of the tree requires updating so that the peeling algorithm does not need to be applied for the complete tree every time a part of a tree changes. The following listing shows where the `CalculationNode`s get updated during the execution of the main loop.

```
1  Store state
2  Propose new state
3  store calculation nodes
4  check dirtyness of calculation nodes (requiresRecalculation())
5  logP = calculateLogPosterior();
6  if (new state is acceptable)
7      accept state
8      mark calculation nodes clean (store())
9  else
10     restore state
11     restore calculation nodes (restore())
```

After a new state is proposed, every `CalculationNode` that may be affected by a changed `StateNode` gets asked to store its internal states in line 3 by calling the `store` method. So, in the model of Figure 12.8, if the kappa parameter changed, the `HKY`, `SiteModel` and `TreeLikelihood` calculation nodes will have their `store` method called. But if only the tree changed but not the kappa parameter, only the `TreeLikelihood` will have its `store` method called, while the `HKY` and `SiteModel` nodes are not disturbed.

Next, in line 4, the same set of `CalculationNodes` will be asked to check whether they are still up to date by calling `requiresRecalculation` on these BEAST-objects. The order in which calculation nodes are called is such that if an input of a `CalculationNode` is a potentially affected `CalculationNode` then the input will have its `requiresRecalculation` method called first. Typically, the method is used to check all inputs that are `StateNodes` and `CalculationNodes` and if one of them is flagged as dirty, a flag is set inside the `CalculationNode` to mark which parts of the calculation needs to be redone to become up to date. Note that this is not yet a good point to actually perform the calculation since it is possible that the calculation is not required. For example, when a state is proposed that violates one of the priors, like a monophyletic constraint on a tree, then it is not necessary to perform the expensive tree likelihood calculation since at the point of checking the prior it is already known that the state is not going to be acceptable.

When a proposed state is acceptable, the set of `CalculationNodes` get their `accept` method called (line 8). This method is used to mark the `CalculationNode` as clean again, and typically no other action is necessary. If the state is not acceptable, the `CalculationNodes` might need to restore their internal state that was stored when the `store` method was called from line 3. This happens in line 11, and the `CalculationNode` should be marked as being clean again. Clearing the dirty flag is done by default in the `store` and `restore` implementations in the `CalculationNode` class, so it is important to call `super.store()` and `super.restore()` from implementations of `store` and `restore`, respectively.

12.3.2   State, stateNodes and initialisation

The state is explicit in XML and as a BEAST-object (unlike BEAST 1). The `State` contains `StateNodes`, and parameters and trees are `StateNode` implementations. A `Parameter` is a `StateNode` for representing a singleton, an array or a matrix of values. A `Tree` is a `StateNode` consisting of `Nodes`. The `State` can store and restore itself for MCMC proposals, which means that `StateNodes` must be able to store and restore their current values. The easiest way to achieve this is by keeping a copy of the values of a `StateNode`, so `Parameters`, for example, contain two arrays, one with the current values and one with stored values. Storing parameter values is implemented as copying the current values in the stored values array, and restoring is implemented as switching stored and current values arrays.

`Operator` BEAST-objects work on the `StateNodes` and when a `StateNode` changes it must report to the `State` that it became dirty. The `State` can then calculate which `CalculationNodes` might be affected by the `StateNode` change. When a

StateNode changes, it is marked as being dirty so that CalculationNodes can interrogate a StateNode and determine whether the CalculationNode needs to recalculate itself.

The State can be stored to disk and later restored from disk, for example for resuming a chain that has not quite converged yet. To save and load a State, the state uses the toXML and fromXML methods of StateNodes.

At the start of a chain, the State and thus its StateNodes need to be initialised. Parameters and trees have default implementations for initialisation, but sometimes there are dependencies between StateNodes, for example in a *BEAST analysis the species tree must contain all gene trees. In such a situation, a complex initialisation is required, which is best done in a StateNodeInitialiser. This is a BEAST-object that has as input one of more StateNodes and can initialise one StateNode based on the values of another. For instance, RandomGeneTree takes as input two trees, one representing the species tree and one representing the gene tree. It initialises the gene tree such that the root of the species tree is lower than the first coalescent event in the gene tree, so all internal nodes of the gene tree are in effect sticking out above the root of the species tree, ensuring the gene tree is consistent with the species tree. State node initialisers are called at the start of the MCMC chain, hence they are input to the MCMC BEAST-object.

A restriction on StateNodes is that none of its inputs can be StateNodes, since that would confuse the State in determining which nodes to update.

12.3.3   CalculationNode and distribution

CalculationNodes are BEAST-objects that actually calculate some things of interest, such as a transition probability matrix, and ultimately result in the calculation of a posterior probability of the state. Writing a CalculationNode can be as simple as implementing the store, restore and requiresRecalculation methods. A Distribution is a CalculationNode that can return a log probability. Distributions are loggable.

CalculationNodes can have StateNodes and other CalculationNodes as input, but there are a few restrictions:

- They are not allowed to form cycles. So, if CalculationNode A is an input of CalculationNode B, then A cannot have B as its input, nor any of the outputs downstream from B.
- All BEAST-objects between a StateNode that is part of the State and the posterior (which is an input of MCMC) must be CalculationNodes. The State notifies all CalculationNodes of changes in StateNodes and BEAST-objects that are not CalculationNodes cannot be notified properly.

12.3.4   Operators

Operators determine how the state space is explored. An Operator has at least one StateNode as input and implements the proposal method. Most operators can be found in the `beast.evolution.operator` package. The following is a list of commonly used operators, a short description and an indication of when it is appropriate to use them.

- A `ScaleOperator` picks a random number s to scale a `StateNode`. If the `StateNode` is a `Parameter`, values of the parameter are multiplied with s, and if it is a `Tree` all internal node heights are multiplied by s. This is the workhorse of the operators for sampling values of real-value parameters, and also useful for scaling complete trees.

- An `UpDownOperator` scales one or more `StateNodes`, but some of them are scaled up (multiplied by s) while others are scaled down (multiplied by $1/s$). This is especially useful for `StateNodes` that are dependent, for example, tree height and population size. Without `UpDownOperator` it is often hard to make such dependent parameters mix. The easiest way to find out whether `StateNodes` are dependent is to plot the pairs of samples (for example with the Tracer program). If the plot shows a cloud of points there is little dependency, but if all points are on a line there is high dependency and the `UpDownOperator` could help to fix this.

- A `RealRandomWalkOperator` (`IntRandomWalkOperator`) selects a random dimension of the real (integer) parameter and perturbs the value by a random amount. This is useful for integer-value parameters.

- A `UniformOperator` is an operator that selects a random dimension of a real or integer parameter and replaces its value by a random value in the range of the parameter. This is useful for sampling integer parameters, but with bolder proposals than the `IntRandomWalkOperator`. A combination of a `UniformOperator` together with a random walk operator usually allows for efficient exploration of the sample space.

- A `BitFlipOperator` selects a random dimension of a boolean parameter and changes its value to false if true, or vice versa. This is useful for boolean-value parameters.

- A `DeltaExchange` operator picks two values in a multi-dimensional parameter and increases one value by a randomly selected value δ, while decreasing the other value by the same amount δ. This is useful in sampling parameter values that are constrained to a certain sum. For example, when sampling a frequencies parameter that represent a substitution's model equilibrium state frequencies, the parameter is restricted to sum to unity.

- A `SwapOperator` swaps one or more pairs of values in a parameter. This is similar to `DeltaExchange`, but especially useful for sampling categories associated with metadata attached to tree branches. An example is rate categories for the uncorrelated relaxed clock model.

- The `Uniform`, `SubtreeSlide`, `Exchange` and `WilsonBalding` operators create proposals for trees, where the first one never changes the topology of a tree, but the others can. The `Exchange` operator comes in a 'narrow' and 'wide' variant for more conservative and more bold moves, respectively. Using a mix of these operators tends to explore the tree space efficiently and makes a chain converge to high posterior trees quickly.

- The NodeReheight operator is a tree operator that randomly changes the height of a node in a tree, then reconstructs the tree from node heights. It has the same function as the set of operators mentioned in the previous paragraph with only a different approach to exploring tree space.
- TipDatesRandomWalker and TipDatesScaler are operators that work on the tip nodes in a tree, that is, the nodes representing the taxa in a tree. As their name suggest, these are useful for estimating tip dates.

Often it is possible to detect that a proposal will surely be rejected, for instance, when the proposal results in a tree with negative branch lengths. To speed up rejection of such a proposal, the proposal method can return negative infinity and the MCMC loop will skip any calculation and rejects immediately.

Gibbs operator: To implement a Gibbs operator, the proposal method can return positive infinity, which ensures the proposal will always be accepted.

12.3.5   Logger and loggable

Implementing loggers is a matter of implementing the Loggable interface, which has three methods: init for initialising the logger, for instance, print the name of the loggable in the header; log which actually logs one or more values; and close for closing off a log. A much used implementation of Loggable is Distribution, which logs the log probability of the distribution, Parameter for logging parameter values and Tree for logging trees in Newick format. Other useful loggers are:

- ESS for reporting the effective sample size of a Function such as a parameter. This is especially useful for tracking the state of a chain when logging to screen.
- MRCATime for reporting the height of an internal node in a tree representing the most recent common ancestor of a set of taxa.
- TreeHeightLogger for reporting the height of the root of a tree.
- TreeWithMetaDataLogger for logging a Newick tree where the branches are annotated with metadata, such as branch rates.

12.4   The evolution library

The evolution library can be found in the beast.evolution package, and contains BEAST-objects for handling alignments, phylogenetic trees and various BEAST-objects for calculating the likelihood of an alignment for a tree and various priors. You can find the details of the individual classes by reading the Java-doc documentation, or by directly looking at the Java classes. In this section, we concentrate on how the various packages and some of the classes inside these packages are related to each other and give a high-level overview of the library.

12.4.1   Alignment, data-type, distance

The data that we want to analyse in evolutionary problems often consist of aligned sequences. The beast.evolution.alignment package contains classes for

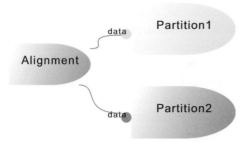

Figure 12.9 Splitting an alignment into two alignments using two filtered alignments, one partition for every third site in the alignment, and one for the first and second sites, skipping every third site. For each of these partitions a likelihood can be defined.

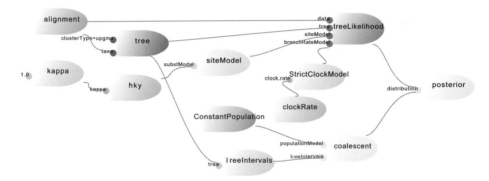

Figure 12.10 Example illustrating most of the components of the evolution library.

handling sequences and alignments. To create a partition of an alignment, an alignment can be filtered using the `FilteredAlignment` class as illustrated in Figure 12.9.

The data in a sequence are of a certain data-type, such as nucleotide or amino acid. The `beast.evolution.datatype` package represents various data-types. To define a new data-type it is possible to use the `UserDataType` class, which only requires changes in the XML. Alternatively, a new data-type can be created by implementing the `DataType` interface, or derived from `DataType.Base`.

It can be useful to calculate the distance between two sequences. The `beast.evolution.alignment.distance` package provides some distances, including Jukes–Cantor distance and Hamming distance. These distances can be used to construct, for example, UPGMA or neighbour-joining trees as starting trees for an MCMC analysis (through the `beast.util.ClusterTree` class).

12.4.2   Tree-likelihood

Figure 12.10 shows a model for a HKY substitution model, and strict clock and co-alescent prior with constant population size that illustrates most of the components of the evolution library. Let's have a look at this model, going from the posterior at the right down to its inputs. The posterior is a compound distribution from the MCMC library. Its inputs are a coalescent prior and a tree-likelihood representing the prior and likelihood for this model. The coalescent is a tree prior with a demographic component and the simple coalescent with constant population size can be found in the `beast.evolution.tree.coalescent` package together with a number of more complex tree priors, such as (extended) Bayesian skyline plot.

The `beast.evolution.likelihood` package contains the tree-likelihood classes. By default, BEAST tries to use the BEAGLE implementation of the peeling algorithm, but otherwise uses a Java implementation. Since the tree-likelihood calculates the likelihood of an alignment for a given tree, it should come as no surprise that the tree-likelihood has an alignment and a tree as its input. The `beast.evolution.alignment` package contains classes for alignments, sequences and taxon sets. The `beast.evolution.tree` package contains the `Tree` state node and classes for initialising trees randomly (`RandomTree`), logging tree information (`TreeHeightLogger`, `TreeWithMetaDataLogger`) and `TreeDistribution`, which is the base class for priors over trees, including the coalescent-based priors.

There is another group of tree priors that are not based on coalescent theory but on theories about speciation, such as the Yule and birth–death priors. Since only a single prior on the tree should be specified, none of these priors is shown in Figure 12.10. These priors can be found in the `beast.evolution.speciation` package together with priors for *BEAST analysis and utility classes for initialisation species trees and logging for *BEAST.

The tree-likelihood requires a site model as input, which takes a substitution model as input, a HKY model in Figure 12.10. In the evolution library, there are packages for site models and substitution models. The site model package (`beast.evolution.sitemodel`) only contains an implementation of the gamma site model, which allows a proportion of the sites to be invariant. The substitution model package (`beast.evolution.substitutionmodel`) contains the most popular models, including Jukes–Cantor, HKY and general time-reversible substitution models for nucleotide data as well as JTT, WAG, MTREV, CPREV, BLOSUM and Dayhoff substitution models for amino-acid data. Root frequencies are in the substitution model package.

The tree-likelihood has a branch-rate model input, which can be used to define clock models on the tree. In Figure 12.10, a strict clock model is shown. The package `beast.evolution.branchratemodel` contains other clock models, such as the uncorrelated relaxed clock model and the random local clock model.

There is a package for operators in the evolution library, which contains most of the `Operator` implementations. It is part of the evolution library since it contains operators on trees, and it is handy to have all general-purpose operators together in a single package. More details on operators can be found in Section 12.3.4.

12.5   Other bits and pieces

There are a few more notable packages that can be useful that are outside the core and evolution packages, namely the following.

The `beast.math` package contains classes mainly for mathematical items. The `beast.math.distributions` package contains distributions for constructing prior distributions over parameters and MRCA times. The `beast.math.statistics` package contains a class for calculating statistics and for entering mathematical calculations.

The `beast.util` package contains utilities such as random number generation, file parsing and managing packages. It contains `Randomize` for random number generation. Note that it is recommended to use the `Randomizer` class for generating random numbers instead of the `java.util.Random` class because it makes debugging a lot easier (see Section 14.5.2) and helps ensuring an analysis started with the same seed leads to reproducible results. The `beast.util` package contains classes for reading and writing a number of file formats, such as `XMLParser` and `XMLProducer` for reading and writing BEAST XML files, and `NexusParser` for reading a subset of NEXUS files. `TreeParser` parses Newick trees. `LogAnalyser` handles trace log files and calculates some statistics on them. Further, the `beast.util` package contains classes for installing, loading and uninstalling BEAST 2 packages.

The `beast.app` package contains applications built on the MCMC and evolution libraries, such as BEAST and BEAUti, and its classes are typically not reused with the exception of input-editors for BEAUti. See Section 15.3.2 for more details.

12.6   Exercise

Open ModelBuilder (java-cp beast.jar beast.app.ModelBuilder). In ModelBuilder, open some XML files from the examples directory and inspect the structure of the graph. It is probably useful to make some BEAST-objects invisible by using the entries in the view menu.

13 BEAST XML

BEAST uses XML as a file format for specifying an analysis. Typically, the XML file is generated through BEAUti, but for new kinds of analysis or analyses not directly supported by BEAUti, it is necessary to construct the XML by hand in a text editor. BEAST-object developers also need to know how to write BEAST XML files in order to test and use their BEAST-objects. BEAUti also uses XML as a file format for specifying templates, which govern its behaviour.

This chapter starts with a short description of XML, then explains how BEAST interprets XML files and how an XML file can be modified. Finally, we work through an example of a typical BEAST specification.

13.1   What is XML?

XML stands for eXtensible Markup Language and has some similarities with HTML. However, XML was designed for encoding data, while HTML was designed for displaying information. The easiest way to explain what XML is without going into unnecessary detail is to have a look at the example shown in Figure 13.1.

Here, we have an (incomplete) BEAST specification that starts with the so-called *XML declaration*, which specifies the character set used (UTF-8 here) and is left unchanged most of the time, unless it is necessary to encode information in another character set. The second line shows a *tag* called 'beast'. Tags come in pairs, an opening tag and a closing tag. Opening tags are of the form `<tag-name>` and can have extra information specified in *attributes*. Attributes are pairs of names and values such as `version='2.0'` for the beast-tag in the example. Names and values are separated by an equal sign and values are surrounded by single or double quotes. Both are valid, but they should match, so a value started with a double quote needs to end with a double quote. Closing tags are of the form `</tag-name>` and have no attributes. Everything between an opening tag and closing tag is called an *element*.

Elements can have other elements nested in them. In the example above, the `input` element is nested inside the `beast` element. Likewise, the `kappa` element is nested inside the `input` element. When an element does not have any other elements nested inside the opening and closing tag can be combined in an abbreviated tag of the form `<tag-name/>`. In the example, the `data` element only has an

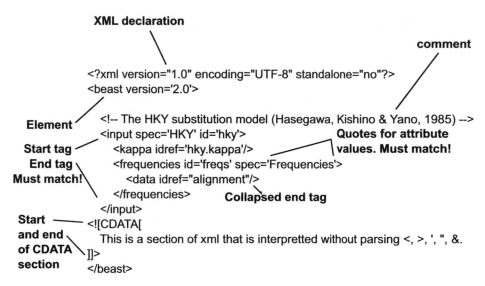

XML declaration

comment

```
<?xml version="1.0" encoding="UTF-8" standalone="no"?>
<beast version='2.0'>
```

Element

Start tag

End tag
Must match!

```
<!-- The HKY substitution model (Hasegawa, Kishino & Yano, 1985) -->
<input spec='HKY' id='hky'>
    <kappa idref='hky.kappa'/>
    <frequencies id='freqs' spec='Frequencies'>
        <data idref="alignment"/>
    </frequencies>
</input>
```

Quotes for attribute values. Must match!

Collapsed end tag

Start and end of CDATA section

```
<![CDATA[
    This is a section of xml that is interpretted without parsing <, >, ', ", &.
]]>
</beast>
```

Figure 13.1 Small XML example with all of its items annotated.

`idref` attribute and has no element enclosed, so it can be abbreviated from `<data idref="alignment"></data>` to `<data idref="alignment"/>`.

XML *comments* start with `<!--` and end with `-->` and text in between is ignored. Any text except double dash `--` is allowed inside comments.

There are a few special characters that should be used with care. Since tags are identified by < and > characters, these cannot be used for attribute names or value or tag names. The other special characters are single and double quote and ampersand. XML *entities* are interpreted as follows:

Table 13.1 XML entities are special combinations of characters starting with & and finishing with ; that are used to include special characters in the text of XML documents

XML entity	Interpreted character
&lt;	<
&gt;	>
&dquot;	"
&quot;	'
&amp;	&

CDATA sections are XML constructs that are interpreted as literal strings, so the content is not parsed. Without the CDATA section, every <, >, ', " and & character would require an XML entity, which would not make such fragments very readable.

Since elements are nested within other elements the nesting defines a hierarchy. So, we can speak of a parent and child relationship between an element and those nested within it. The first element is not nested inside any other element and is called the top-level element. Only one top-level element is allowed.

13.2   BEAST file format and the parser processing model

Since everything in BEAST is a BEAST-object, connected to other BEAST-objects by inputs, a natural way to interpret an XML file is to create a BEAST-object for every XML element (with a few exceptions, such as construction elements like maps and plates and references to other object through the `idref` mechanism). Every attribute and nested element are then inputs to this parent BEAST-object. BEAST considers an XML file as a specification for doing an analysis, so it will first determine the kind of analysis by looking for a BEAST-object that is runnable (that is, a BEAST-object like MCMC that extends `beast.core.Runnable`) among the children of the BEAST element.

13.2.1   Reserved XML attributes

There are four attribute names that are reserved and have special meaning: `id`, `idref`, `name` and `spec`. The tag `input` can be used for every BEAST-object. The `name` attribute specifies the name of the input that connects the BEAST-object to its parent. So, the `name` attribute is actually a property of the parent BEAST-object that the XML specifies. The `spec` attribute specifies the Java class of the BEAST-object. The `id` attribute specifies a unique identifier for a BEAST-object. BEAST-objects can be connected to several inputs using the `id`/`idref` mechanism; the `id` of a BEAST-object can be referred to from another element by specifying an `idref` attribute with the value of an `id`. Each BEAST-object has a unique `id`, but the `name`, `spec` and `idref` attributes need not be unique in an XML file. Some inputs are not BEAST-objects but primitives (namely Integer, Double, Boolean, or String) and these inputs can be specified through an attribute. The following XML fragment illustrates these concepts.

```
1 <input name="operator" id="kappaScaler"
2     spec="beast.evolution.operators.ScaleOperator"
3     scaleFactor="0.5" weight="1">
4     <input name="parameter" idref="hky.kappa"/>
5 </input>
```

There are two `input` elements, the first one specifying a BEAST-object of class `beast.evolution.operators.ScaleOperator`. The attributes `scaleFactor` and `weight` set primitive inputs to value 0.5 and 1, respectively. The scale operator has an input with name *parameter*, and the nested element refers through the `idref` attribute to a BEAST-object that should be defined elsewhere in the XML.

13.2.2   Name spaces

The above listing looks quite verbose, and there are a few mechanisms in BEAST to make the XML shorter and more readable. The Java class names of BEAST-object can become quite long, and to increase readability the parser recognises *name spaces*; a list of name spaces is specified by the `namespace` attribute on the top-level BEAST element. The list is a colon-separated list of packages; for example, to specify a name space containing `beast.core` and `beast.evolution.operators` use

```
1  <beast namespace="beast.core:beast.evolution.operators">
```

The parser finds a BEAST-object class by going through the list and appending the value of the `spec` attribute to the package name. By default, the top-level package is part of the list (at the end), even when not specified explicitly in the `namespace` attribute. With the above name space definition, the fragment shown earlier is equivalent to

```
1  <input name="operator" id="kappaScaler"
2     spec="ScaleOperator" scaleFactor="0.5" weight="1">
3     <input name="parameter" idref="hky.kappa"/>
4  </input>
```

Note that if there are BEAST-objects with the same name in different packages, the BEAST-object that matches with the first package in the class path is used. To prevent such name clashes, the complete class name can be used.

13.2.3   Input names

Using `input` elements as described above, the name of the input is explicitly defined in a `name` attribute and the tag name `input` is constant and rather uninformative. As an alternative, BEAST recognises elements in which the input name is used as the tag name of a child element. That is, `<input name="xyz"/>` can also be encoded as `<xyz/>`. This allows us to write the fragment above as

```
1  <operator id="kappaScaler" spec="ScaleOperator"
2     scaleFactor="0.5" weight="1">
3     <parameter idref="hky.kappa"/>
4  </operator>
```

Note that when the `name` attribute is specified, the tag name will be ignored. Further, the end tag should have the same name as the start tag, so both have the tag name 'operator'.

13.2.4   Referring to elements with id/idref mechanism

The `id`/`idref` mechanism is used extensively throughout BEAST XML and can take up quite a lot of text. When an `idref` attribute is specified, all other attributes (except the name attribute) will be ignored, so `<parameter spec="HKY" idref="hky.kappa"/>` is interpreted the same as `<parameter idref="hky.kappa"/>` and the `spec` attribute is ignored. This pattern is very common, and to shorten the XML the parser recognises an attribute that has a value starting with @ as an element with a single `idref` attribute. That is, attribute `parameter="@hky.kappa"` will be recognised as a nested element `<parameter idref="hky.kappa"/>`. With this in mind, we see that the following fragment is equivalent to that at the start of this section.

```
1  <operator id="kappaScaler" spec="ScaleOperator"
2    scaleFactor="0.5" weight="1" parameter="@hky.kappa"/>
```

Table 13.2 Reserved element names

Tag name	Associated BEAST-object
run	Must be `beast.core.Runnable`
distribution	Must be `beast.core.Distribution`
operator	Must be `beast.core.Operator`
logger	Is `beast.core.Logger`
data	Is `beast.evolution.alignment.Alignment`
sequence	Is `beast.evolution.alignment.Sequence`
state	Is `beast.core.State`
parameter	Is `beast.core.parameter.RealParameter`
tree	Is `beast.evolution.tree.Tree`
input	Reserved name
map	See Section 13.2.6
plate	Macro, see Section 13.2.5
mergepoint	Reserved for BEAUti templates
mergewith	Reserved for BEAUti templates

13.2.5   Plates for repetitive parts

Especially for multi-gene analysis, XML files can contain parts that are the same but differ only in the name of a gene. To compress such parts of an XML file, the XML parser pre-processes an XML file by looking for plate elements. You can look at a plate as a kind of macro for a loop over a part of the XML. A plate has two attributes, `var` which defines the name of the variable and `range` which is a comma-separated list of values assigned to the variable name. The plate is replaced by the XML inside the plate copied once for each of the values, and wherever the variable name occurs this is replaced by the value. Variable names are encoded as `$(var)`. For example, the following fragment

```
1  <plate var="n" range="red,green,blue">
2          <parameter id="color.$(n)" value="1.0"/>
3  </plate>
```

is interpreted as

```
1          <parameter id="color.red" value="1.0"/>
2          <parameter id="color.green" value="1.0"/>
3          <parameter id="color.blue" value="1.0"/>
```

Note that plates can be nested when different variable names are used.

13.2.6   Element names

Some BEAST-objects are so common that the `spec` attribute can be omitted and the parser still knows which BEAST-object to use. Table 13.2 lists the element names and associated BEAST-objects. Note that if the `spec` attribute is used then the tag name is ignored for determining the class of a BEAST-object. The parser can be told to extend the mapping of element names to BEAST-object classes using a map-element. The map element has a `name` attribute that defines the tag name and the text content describes the class. For example, to map tag name `prior` to `beast.math.distributions.Prior`, use

```
1 <map name='prior'>beast.math.distributions.Prior</map>
```

And for elements with name `prior` a `beast.math.distributions.Prior` BEAST-object will be created. Note that name spaces are used, so if `beast.math.distributions` is a name space, only `Prior` needs to be used inside the map element. These map elements must be children of the top-level BEAST element.

13.2.7 XML parsing summary

A BEAST XML file will always contain an analysis that starts at the first child of the top-level element that is a runnable BEAST-object. The analysis is constructed by creating BEAST-objects and setting input values of BEAST-objects, possibly connecting them with other BEAST-objects.

The *class* of a BEAST-object for an element is determined as follows:

- If an `idref` attribute is specified, then the class of the BEAST-object for the element with the `id` of the same name is used, e.g. `<input idref="hky.kappa"/>`.
- If a `spec` attribute is specified, then the first name space containing the spec-value is used, e.g. `<operator spec="ScaleOperator"/>`.
- If the element name is defined in a map element, then the class defined in the map element is used, e.g. `<scaleoperator/>` when `<map name="scaleoperator"> beast.evolution.operators.Scale Operator </map>` is defined.
- If the element name is a reserved name, then the BEAST-object listed in Table 13.2 is used, e.g. `<parameter value="3.14"/>`.

The *input name* associated with a BEAST-object is determined as follows:

- If the `name` attribute is specified in a child element, the name's value is used, for example, `<input name="operator">`.
- If the child element name differs from input, then the element name is used, for example, `<operator>`.
- If the tag is 'input' and contains text, then the input name 'value' is used, for example, `<input>3</input>`.

Finally, *the input* value of a BEAST-object is determined as follows:

- If an `idref` is specified, then the referred object is used as value, e.g. `<parameter idref="hky.kappa">` or `parameter="@hky.kappa"`.
- If an attribute value is specified, then the primitive value is used, e.g. `weight="3.0"`.
- If there is text content inside the element, then the text value is used, e.g. `<input>3.0</input>`.

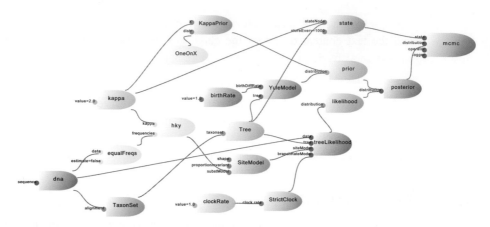

Figure 13.2 Model view of the annotated XML example.

13.3   An annotated example

The following analysis estimates the tree using an HKY substitution model for which
the kappa is estimated and a strict clock model. A Yule prior is placed on the tree and a
1/X prior on kappa. Comments are added to highlight peculiarities of the BEAST XML
parser. Figure 13.2 shows the model view of the file, but with sequences and loggers
removed for clarity.

```
1 <?xml version="1.0" encoding="UTF-8" standalone="no"?>
2 <beast version="2.0" namespace="beast.core:beast.evolution.
    alignment:beast.evolution.tree:beast.evolution.tree.
    coalescent:beast.core.util:beast.evolution.nuc:beast.evolution.
    operators:beast.evolution.sitemodel:beast.evolution.
    substitutionmodel:beast.evolution.likelihood:beast.math.
    distributions:beast.evolution.speciation:beast.evolution.
    branchratemodel">
```

The first line is the XML declaration which tells the parser about the character encoding. The second line indicates that this is a BEAST version 2 file. Furthermore, a list of
packages is defined that constitute the name space.

```
3 <data dataType="nucleotide" id="dna">
4     <sequence taxon="human">
        AGAAATATGTCTGATAAAAGAGTTACTTTGATAGAGTAAATAATAGGAGC...</
        sequence>
5     <sequence taxon="chimp">
        AGAAATATGTCTGATAAAAGAATTACTTTGATAGAGTAAATAATAGGAGT...</
        sequence>
6     <sequence taxon="bonobo">
        AGAAATATGTCTGATAAAAGAATTACTTTGATAGAGTAAATAATAGGAGT...</
        sequence>
7     <sequence taxon="gorilla">
        AGAAATATGTCTGATAAAAGAGTTACTTTGATAGAGTAAATAATAGAGGT...</
        sequence>
```

```
8    <sequence taxon="orangutan">
         AGAAATATGTCTGACAAAAGAGTTACTTTGATAGAGTAAAAAATAGAGGT...</
         sequence>
9    <sequence taxon="siamang">
         AGAAATACGTCTGACGAAAGAGTTACTTTGATAGAGTAAATAACAGGGGT...</
         sequence>
10 </data>
```

Note that `data` (line 3) and `sequence` (lines 4 to 9) are reserved names that
are associated with `beast.evolution.alignment.Alignment` and `beast.`
`evolution.alignment.Sequence` BEAST-objects, respectively. Most BEAST
XML files have a data block element at the start. The sequence BEAST-object has
an input called 'value' and the text inside the sequence tags representing character
sequences are assigned to this 'value' input. The dots in the XML fragment indicate
that there are a lot more sites in the sequence not shown here to save space. The data
element has an `id` attribute, so that it can later be referred from, for example, the tree-
likelihood.

```
11 <distribution id="posterior" spec="util.CompoundDistribution">
12     <distribution id="prior" spec="util.CompoundDistribution">
13         <distribution conditionalOnRoot="false" id="YuleModel"
               spec="YuleModel" tree="@Tree">
14             <parameter estimate="false" id="birthRate"
15 name="birthDiffRate" value="1.0"/>
16         </distribution>
17
18         <distribution id="KappaPrior" spec="Prior" x="@kappa">
19             <distr id="OneOnX.0" spec="OneOnX"/>
20         </distribution>
21     </distribution>
22
23     <distribution data="@dna" id="likelihood" spec="TreeLikelihood"
           tree="@Tree">
24         <siteModel id="SiteModel" spec="SiteModel">
25             <substModel id="hky" kappa="@kappa" spec="HKY">
26                 <frequencies data="@dna" estimate="false" id="
                       equalFreqs" spec="Frequencies"/>
27             </substModel>
28         </siteModel>
29
30         <branchRateModel id="StrictClock" spec="StrictClockModel">
31             <parameter estimate="false" id="clockRate" name="clock.
                   rate" value="1.0"/>
32         </branchRateModel>
33     </distribution>
34 </distribution>
```

Lines 11 to 34 define the posterior from which we sample. Note that the `spec`
attribute in line 11 contains part of the package (`util.CompoundDistribution`)
that contains the `CompoundDistribution` class. The posterior contains two distributions: a compound distribution for the prior (line 12); and a tree-likelihood for the
likelihood (line 23). The prior consists of a Yule prior (line 13) and a prior on kappa

(line 18). Note that the kappa parameter referred to in line 18 is defined in the state (line 37), showing that idrefs can refer to BEAST-objects specified later in the file.

The tree likelihood has a sitemodel as input (line 24), which here has a HKY substitution model (line 25) as input. The likelihood also has a clock model (line 30) which here is a strict clock. All elements have `id` attributes so that they can be referred to, for example, for logging.

Lines 35 to 78 specify the MCMC BEAST-object. This is the main entry point for the analysis. The child elements of the MCMC element are the state, the distribution to sample from, a list of operators, a list of loggers and an initialiser for the tree. The state (line 37 to 44) lists the state nodes that are operated on, here just the tree and kappa parameter. Note that the tree element (line 38) has a `name` attribute linking it to the state through its `stateNode` input. If the tag name would be set to `stateNode`, the `name` attribute is superfluous, but a `spec` attribute is required to specify the class, which is implicit when 'tree' is used as a tag.

```
35  <run chainLength="10000000" id="mcmc" preBurnin="0" spec="MCMC">
36
37      <state id="state" storeEvery="100000">
38          <tree estimate="true" id="Tree" name="stateNode">
39              <taxonset id="TaxonSet" spec="TaxonSet">
40                  <data idref="dna" name="alignment"/>
41              </taxonset>
42          </tree>
43          <parameter estimate="true" id="kappa" lower="0.0"
                name="stateNode" value="1.0"/>
44      </state>
45
46      <distribution idref="posterior"/>
```

The distribution (line 46) the MCMC analysis samples from refers to the posterior defined earlier in the file at line 11. Lines 47 to 53 list the operators used in the MCMC chain. Operators need to refer to at least one state node defined in the state element (line 37 to 44).

```
47  <operator degreesOfFreedom="1" id="treeScaler" scaleFactor="0.5"
        spec="ScaleOperator" tree="@Tree" weight="1.0"/>
48  <operator id="UniformOperator" spec="Uniform" tree="@Tree"
        weight="10.0"/>
49  <operator gaussian="true" id="SubtreeSlide" optimise="true" size="1.0"
        spec="SubtreeSlide" tree="@Tree" weight="5.0"/>
50  <operator id="narrow" isNarrow="true" spec="Exchange" tree="@Tree"
        weight="1.0"/>
51  <operator id="wide" isNarrow="false" spec="Exchange" tree="@Tree"
        weight="1.0"/>
52  <operator id="WilsonBalding" spec="WilsonBalding" tree="@Tree"
        weight="1.0"/>
53  <operator degreesOfFreedom="1" id="KappaScaler" scaleFactor="0.5"
        spec="ScaleOperator" weight="1.0" parameter='@kappa'/>
```

To log the states of the chain at regular intervals, three loggers are defined; a trace logger (line 54), which can be analysed by the Tracer program; a screen logger (line 63), which gives feedback on screen while running the chain; and a tree logger (70) for writing a NEXUS file to store a tree set.

```
54 <logger fileName="beast.$(seed).log" id="tracelog" logEvery="10000"
       model="@posterior">
55     <log idref="posterior"/>
56     <log idref="likelihood"/>
57     <log idref="prior"/>
58     <log id="TreeHeight" spec="TreeHeightLogger" tree="@Tree"/>
59     <log idref="YuleModel"/>
60     <log idref="kappa"/>
61 </logger>
62
63 <logger id="screenlog" logEvery="10000">
64     <log idref="posterior"/>
65     <log arg="@posterior" id="ESS.0" spec="util.ESS"/>
66     <log idref="likelihood"/>
67     <log idref="prior"/>
68 </logger>
69
70 <logger fileName="beast.$(seed).trees" id="treelog" logEvery="10000"
       mode="tree">
71     <log idref="Tree"/>
72 </logger>
```

The last input for the MCMC BEAST-object is a tree initialiser:

```
73 <init id="RandomTree" initial="@Tree" spec="RandomTree" taxa="@dna">
74     <populationModel id="ConstantPopulation"
           spec="ConstantPopulation">
75         <parameter dimension="1" estimate="true" id="popSize"
               name="popSize" value="1"/>
76     </populationModel>
77 </init>
```

Finally, the XML file needs a closing tag for the run element and a closing element for the top-level BEAST element.

```
78 </run>
79 </beast>
```

A notable difference from BEAST 1 is that the order in which BEAST-objects are specified does not matter.

13.4   Exercise

Are the following XML fragments equivalent, assuming that the name space is 'beast.evolution.sitemodel: beast.evolution. substitutionmodel: beast.evolution.likelihood'?

Fragment 1

```
1 <input name='substModel' id="hky" spec="HKY">
2     <input name='kappa' idref="hky.kappa" >
3     <input name='frequencies' id="freqs" spec="Frequencies">
4             <input name='data' idref="alignment"/>
5     </input>
6 </input>
```

```
 7
 8 <input spec="TreeLikelihood">
 9     <input name='data' idref='alignment'/>
10     <input name='tree' idref='tree'/>
11     <input name='siteModel' spec="SiteModel">
12         <input name='substModel' idref='hky'/>
13     </input>
14 </input>
```

Fragment 2

```
1 <substModel id="hky" spec="HKY" kappa="@hky.kappa" >
2     <frequencies id="freqs" spec="Frequencies"
3         data="@alignment"/>
4 </substModel>
5
6 <distribution data="@alignment" spec="TreeLikelihood"
7     tree="@tree">
8     <siteModel spec="SiteModel" substModel='@hky'/>
9 </distribution>
```

14 Coding and design patterns

We will show a few patterns commonly used with BEAST 2, illustrating how the MCMC framework can be exploited to write efficient models. The way BEAST 2 classes are based on BEAST-objects has the advantage that it does automatically define a fairly readable form of XML and allows automatic checking of a number of validation rules. The validation helps in debugging models. These advantages are based on Java introspection and there are a few peculiarities due to the limitations of Java introspection which can lead to unexpected behaviour. These issues will be highlighted in this chapter.

We start with some basic patterns for BEAST-objects, inputs and the core `StateNode` and `CalulationNode` BEAST-objects, and how to write efficient versions of them. We also show a number of the most commonly used BEAST-objects from the evolution library and how to extend them. Finally, there are some tips and a list of common errors that one should be aware of.

14.1 Basic patterns

First, we have a look at some basic patterns.

14.1.1 Basic BEAST-object layout

Box 14.1 shows the layout of most BEAST-objects that you will find in the BEAST code. We all love to have documented code, so it starts with a `Description` annotation, typically a one-line comment on the function of the BEAST-object, but multi-line descriptions are fine as well. The `Description` is used in documentation and in online help in GUIs like BEAUti. Omitting such a description leads to failure of the `DocumentationTest`. This does not mean that further Java comments are not allowed or necessary, but the description does help in documenting classes for code writers as well.

Every BEAST-object implements `BEASTInterface`. `BEASTObject` is a base implementation of `BEASTInterface` and many BEASTObjects extend the `BEASTObject` class.[1] By implementing `BEASTInterface`, services that use

---

[1] Note we use BEAST-object and BEASTObject for implementations of `BEASTInterface` and extensions of `BEASTObject`, respectively.

195

introspection, like validation of models, are automatically provided. To specify an input for a BEAST-object, just declare an `Input` member. Input is a template class, so the type of input can be specified to make sure that when Inputs are connected to BEAST-objects the correct type of BEAST-object is used. At least two strings are used in the constructor of an Input:

- a name of the input, used in the XML, in documentation and in GUIs,
- a description of the input, used in documentation and GUI help.

Other constructors exists to support validation, default values, lists of values, enumerations of Strings, etc. (see Section 14.2 for details).

It is handy to have all inputs together and in the BEAST code they can normally be found at the start of a class. Next in Box 14.1 follow member objects. Often, these are objects that shadow inputs, so that the `Input.get` method does not need to be called every time an input object is accessed. For example, a BEAST-object with input `Input<RealParameter> parameterInput` can be shadowed by a `RealParameter parameter` member object. In the `initAndValidate` method, the parameter is then initialised using `parameter = parameterInput. get();`.

The first method in a BEAST-object is typically the `initAndValidate` method. This serves as a place to perform validation on the Inputs, for instance range checks or checks that dimensions of two inputs are compatible. Furthermore, it is a place to perform everything that normally goes into a constructor. BEAST-objects are typically created by the `XMLParser`, which first sets values for all inputs, then calls `initAndValidate`. Since, Java invokes a constructor before any input can be assigned a value, a separate method is required to do the initialisation after inputs are

Box 14.1 Basic layout of most BEAST-objects in BEAST.

```
1  @Description("Some sensible description of the BEAST-object")
2  public class MyBEASTObject extends BEASTObject {
3      // inputs first
4      public Input<RealParamater> myInput = new Input<>...;
5
6      // members next
7      private Object myObject;
8
9      // initAndValidate
10     @Override
11     public void initAndValidate() {
12       //...
13     }
14
15     // class specific methods
16
17     // Overriding methods
18  }
```

Box 14.2 Layout of the HKY BEAST-object in BEAST.

```
 1 @Description("HKY85 (Hasegawa, Kishino & Yano, 1985) "+
 2         "substitution model of nucleotide evolution.")
 3 @Citation("Hasegawa, M., Kishino, H and Yano, T. 1985. "+
 4         "Dating the human-ape splitting by a "+
 5         "molecular clock of mitochondrial DNA. " +
 6         "Journal of Molecular Evolution 22:160-174.")
 7 public final class HKY extends SubstitutionModel.Base {
 8     public Input<RealParameter> kappa = new Input<RealParameter>
         ("kappa",
 9         "kappa parameter in HKY model", Validate.REQUIRED);
10
11     @Override
12     public void initAndValidate() throws Exception {...}
13
14     @Override
15     public void getTransitionProbabilities(double distance,
         double[] matrix) {...}
16
17     @Override
18     protected boolean requiresRecalculation() {...}
19
20     @Override
21     protected void store() {...}
22
23     @Override
24     protected void restore() {...}
25 }
```

assigned value, for instance, through the XMLParser. So, BEAST-objects do not have a constructor typically (though see Section 14.2.3 for an exception).

After the initialisation, class-specific methods and overriding methods (with notably store, restore and requiresRecalculation at the end) conclude a BEAST-object.

Box 14.2 shows the skeleton of a larger example. Note that apart from the Description annotation, there is also a Citation annotation that can be used to list a reference and DOI of a publication. At the start of a run, BEAST visits all BEAST-objects in a model and lists the citations, making it easy for users to reference work done by BEAST-object developers.

The HKY BEAST-object has a single input for the kappa parameter. There are more details on Input construction and validation in Section 14.2. There is no further validation required in the initAndValidate method, where only a few shadow parameters are initialised. The getTransitionProbabilities method is where the work for a substitution model takes place. The methods requiresRecalculation, store and restore complete the BEAST-object with implementation of CalculationNode methods.

14.2   Input patterns

Inputs can be created for primitives, BEAST-objects, lists or enumerations. However, inputs cannot be template classes, with the exception of `Lists`, due to Java introspection limitations, unless you explicitly provide the class as argument to an Input constructor. By calling the appropriate constructor, the `XMLParser` validates the input after assigning values and can check whether a `REQUIRED` input is assigned a value, or whether two inputs that are restricted by a exclusive-or (`XOR`) rule have exactly one input specified.

14.2.1   Input creation

Inputs can be simple primitives, like Double, Integer, Boolean, String. To create a primitive input, use the `Input<Primitive>` constructor, for example a `Boolean` input can be created like this.

```
1 public Input<Boolean> scaleAllInput =
2     new Input<Boolean>("scaleAll",
3         "if true, all elements of a parameter (not tree) are scaled,
            otherwise one is randomly selected",
4         new Boolean(false));
```

Note at least two arguments are required for an Input constructor: the name of the input and a short description of the function of the input. Inputs of a BEAST-object can be other BEAST-objects, which can be created similarly like this.

```
1 public Input<Frequencies> freqsInput =
2     new Input<Frequencies>("frequencies",
3     "substitution model equilibrium state frequencies");
```

Inputs can have multiple values. When a list of inputs is specified, the Input constructor should contain a (typically empty) List as a start value so that the type of the list can be determined through Java introspection (as far as we know this cannot be done from the declaration alone due to Java introspection limitations).

```
1 public Input<List<RealParameter>> parametersInput =
2     new Input<List<RealParameter>>("parameter",
3         "parameter, part of the state",
4         new ArrayList<RealParameter>());
```

To provide an enumeration as input, the following constructor can be used: it takes the usual name and description arguments, then the default value and an array of strings to choose from. During validation it is checked that the value assigned is in the list.

```
1 public enum LOGMODE {autodetect, compound, tree}
2
3 public Input<LOGMODE> modeInput = new Input<LOGMODE>("mode",
4     "logging mode, one of " + LOGMODE.values(),
5     LOGMODE.autodetect, LOGMODE.values());
```

14.2.2   Input rules

To provide some basic validation, an extra argument can be provided to the Input constructor. By default, inputs are considered to be OPTIONAL, that is, need not necessarily be specified. If an input is a REQUIRED input, this validation rule can be added as an argument to the constructor of the input as follows:

```
1 public Input<Parameter> kappaInput =
2     new Input<Parameter>("kappa",
3         "kappa parameter in HKY model",
4         Validate.REQUIRED);
```

When the XMLParser processes an XML fragment, these validation rules are automatically checked. So, when the kappa input is not specified in the XML, the parser throws an exception. These input rules are also used in BEAUti to make sure the model is consistent, and in documentation generated for BEAST-objects.

If a list of inputs need to have at least one element specified, the required argument needs to be provided.

```
1 public Input<List<Operator>> operatorsInput =
2   new Input<List<Operator>>("operator",
3     "operator for generating proposals in MCMC state space",
4     new ArrayList<Operator>(), Validate.REQUIRED);
```

Sometimes either one or another input is required, but not both. In that case an input is declared XOR and the *other* input is provided as an extra argument. The XOR goes on the second input. Note that the order of inputs matters since at the time of construction of an object the members are created in order of declaration. This means that the first input cannot access the second input at the time just after it was created. Therefore, the XOR rule needs to be put on the second input.

```
1 public Input<Tree> treeInput =
2     new Input<Tree>("tree",
3         "if specified, all tree branch length are scaled");
4 public Input<Parameter> parameterInput =
5     new Input<Parameter>("parameter",
6         "if specified, this parameter is scaled",
7         Validate.XOR, treeInput);
```

14.2.3   Input rule of base class is not what you want

Suppose an Input is REQUIRED for a base class you want to override, but for the derived class this Input can be OPTIONAL. The way to solve this is to set the Input to OPTIONAL in the constructor of the derived class. For example, the GTR BEAST-object represents the GTR substitution model for nucleotides and has six individual inputs for rates. GTR derives from GeneralSubstitutionModel, and its rates are not used. For GeneralSubstitutionModel the rates input is required, but for the GTR BEAST-object, it is ignored. Changing the validation rule for an input is done in the constructor.

```
1    public GTR() {
2        rates.setRule(Validate.OPTIONAL);
3    }
```

Note that the constructor needs to be public to prevent `IllegalAccess Exceptions` on construction by, e.g. the `XMLParser`.

14.2.4   Access inputs for reading, *not writing!*

BEAST-object inputs should typically only be accessed to get values from, not assign values to. For a `RealParameter` input named `input`, use `input.get()` to get the parameter of the input. To get the value of the `RealParameter`, use `input.get().getValue()`. A common pattern is shadowing the input so that the get methods do not need to be accessed every time.

```
RealParamater p;
...
// in initandValidate
p = input.get();
...
// in methods doing calculations
double value = p.getValue();
```

The input only represents the link between BEAST-objects. Say, an input `inputX` has a BEAST-object `X` as its value. By 'writing' an input to a new BEAST-object `Y` a new link is created to `Y`, but that does not replace the BEAST-object `X`. This is not a problem when `X` has no other outputs than `inputX`, but it can lead to unexpected results when there are more outputs. There are exceptions, for example programs for editing models, like BEAUti and ModelBuilder, but care must be taken when assigning input values.

14.3   `initAndValidate` **patterns**

The `initAndValidate` method partly has the function of a constructor, so is used for initialising internal variables such as variables shadowing inputs. The other task is to ensure that the combination of input values is reasonable, and if not, change the input values or throw an exception.

14.3.1   Input parameter dimension is unknown ...

A common situation is where the dimension of a parameter is easy to calculate, but is a nuisance for the user to define. For example, the number of categories in the relaxed clock model should be equal to the number of branches in a tree. The `initAndValidate` method is the ideal place to calculate parameter dimensions and assign it to the parameter using a call `parameter.setDimension(dim)`.

14.3.2 Input parameter *value* is unknown ...

Another common situation is where it is easy to determine initial values for parameters. The categories in the relaxed clock model are initialised as $0, 1, \ldots, n - 1$ where n is the dimension of the categories parameter. For large n, users will find it annoying to get this correct in the XML. To do this in the initAndValidate() method, create a new parameter X and use input.get().assignFromWithoutID(X) to initialise values.

```
1 @Override
2 public void initAndValidate() throws Exception {
3     // determine dimension, number of Nodes in a tree minus 1 here
4     int categoryCount = tree.get().getNodeCount() - 1;
5
6     // initialise array with pre-calculated values
7     Integer [] categories = new Integer[categoryCount];
8     for (int k = 0; k < categoryCount; k++) {
9         categories[k] = k;
10    }
11
12    // create new Parameter with new values
13    IntegerParameter other = new IntegerParameter(categories);
14    categories.assignFromWithoutID(other);
15 }
```

14.4 CalculationNode **patterns**

CalculationNodes are classes jointly responsible for efficiently calculating values of interest, in particular the posterior.

14.4.1 Using requiresRecalculation

The requiresRecalculation method in a CalculationNode is called in the MCMC loop just after a new proposal is made by an operator (see Section 12.3.1). The task of the method is to check whether any of its inputs changed that might affect the internal state of the CalculationNode. There are two potential sources that could have changed; inputs of StateNodes and inputs of other CalculationNodes. To check whether a StateNode is dirty, call somethingIsDirty(). For CalculationNodes, call isDirtyCalculation.

Note, since a StateNode is a CalculationNode, you should test that a class is a StateNode before testing it is a CalculationNode.

```
1 public boolean requiresRecalculation() {
2     // for StateNode inputs only
3     if (stateNodeInput.get().somethingIsDirty()) {
4         return true;
5     }
6
7     // for CalculationNode inputs only
```

```
 8   if (calculationNodeInput.get().isDirtyCalculation()) {
 9       return true;
10   }
11   return false;
12 }
```

14.4.2 Lean CalculationNode

The difference between a lean and a fat CalculationNode is that a lean one
does not store intermediate results, while a fat one does. As a consequence, a lean
CalculationNode needs to recalculate its internal state after a call to restore,
while the fat CalculationNode just reverts back to the previously stored values.
The choice between lean and fat is determined by the amount of work done in an
update. If the calculation is relatively expensive, choose a fat CalculationNode,
but for simple calculations a lean CalculationNode is simpler and does not hurt
performance by much.

To manage a lean CalculationNode, add a flag to indicate whether the internal
state is up to date (needsUpdate in the following fragment). At time of
initialisation, set the flag to true. When a calculation is done, reset the flag. The
requiresRecalculation method should only set the flag when any of its inputs
are dirty. Note that the store method does not change the flag, but restore does.

```
 1 // flag to indicate internal state is up to date
 2 boolean needsUpdate;
 3 // the value to calculate
 4 Object someThing;
 5
 6 public void initAndValidate() {needsUpdate = true;}
 7
 8 // CalculationNode specific interface that returns results
 9 public Object calculateSomeThing() {
10     if (needsUpdate) {
11         update();
12     }
13     return someThing;
14 }
15
16 void update() {
17     someThing = ...;
18     needsUpdate = false;
19 }
20
21 public boolean requiresRecalculation() {
22     if (someInputIsDirty()) {
23         needsUpdate = true;
24         return true;
25     }
26     return false;
27 }
28
29 public void store() {super.store();}
30
```

```
31 public void restore() {
32     needsUpdate = true;
33     super.restore();
34 }
```

14.4.3 Fat `CalculationNode`

A fat `CalculationNode` stores intermediate results of a calculation. So, apart from the result itself it has to reserve memory for the stored results in the `initAndValidate` method. When the `store` method is called, the intermediate results need to be copied to the stored results objects. Restoring amounts to simply swapping references between intermediate and stored results.

```
1 Object intermediateResult;
2 Object storedIntermediateResult;
3
4 public void initAndValidate() {
5     // reserve space for result objects
6     intermediateResult = new ...;
7     storedIntermediateResult = new ...;
8 }
9
10 // CalculationNode specific interface that returns results
11 public Object calculateSomeThing() {
12     intermediateResult = calcIntermediateResult()
13     return calcResult(intermediateResult);;
14 }
15
16 public void store() {
17     // copy intermediateResult to storedIntermediateResult
18     ...
19     super.store();
20 }
21
22 public void restore() {
23     Object tmp = intermediateResult;
24     intermediateResult = storedIntermediateResult;
25     storedIntermediateResult = tmp;
26     super.restore();
27 }
```

14.5 Common extensions

There are a few classes that we would like to highlight for extensions and point out a few notes and hints on how to do this. To add a clock model, implement the `BranchRateModel` interface, which has just one method `getRateForBranch`. To add a new Tree prior, extend `TreeDistribution` (not just `Distribution`) and implement `calculateLogP`.

14.5.1   Adding a substitution model

Extend `SubstitutionModel.Base` class to add a new substitution model. A substitution model should implement the `getTransitionProbabilities` method, which returns the transition matrix for a branch in the tree. `SubstitutionModel` is a `CalculationNode`, so it may be worth implementing it as a fat `CalculationNode`.

14.5.2   Adding an operator

To create a new operator, extend the `Operator` class. An operator should have at least one input with a `StateNode` to operate on and it should implement the `proposal` method, which changes the State. The `proposal` method should return the Hastings ratio. However, it should return `Double.NEGATIVE_INFINITY` if the proposal is invalid, not throw an `Exception`. For Gibbs operators, return `Double.POSITIVE_INFINITY` to ensure the proposal will always be accepted. Consider implementing `optimize()` if there is a parameter of the operator, such as a scale factor or window size. The chain will attempt to change its value, which can greatly help in getting a better acceptance rate and thus better mixing of the chain.

Note, when using random numbers, instead of the standard `java.util.Random` class use the `beast.util.Randomizer` class. This class uses the Mersenne twister algorithm for efficiently generating random numbers that are 'more random' than those produced by the `java.util.Random` class. Further, it is initialised using the seed given to the `BeastMain` class. Therefore, restarting BEAST with the same seed will reproduce the exact same set of random numbers, which makes it easier to debug operators.

14.5.3   Adding a logger

To add a logger, simply implement the `Loggable` interface, which has three methods: `init`, `log` and `close`. There are two main forms of loggers in BEAST, namely trace logs and tree logs. A trace log is a tab-separated file with one log entry per line and a header listing the columns of the log entries. To create a logger that fits in a trace log, the `init` method should print the tab-separated headers and the `log` methods should print tab-separated values (`close` can be left empty). Note that the number of entries in the header should match the number of entries added by the `log` method. Tree logs are produced in NEXUS format.

14.6   Tips

14.6.1   Debugging

Debugging MCMC chains is a hazardous task. To help check that the model is valid, BEAST recalculates the posterior the first number of steps for every third sample.

Before recalculating the posterior, all `StateNodes` become marked dirty so all `CalculationNodes` should update themselves. If the recalculated posterior differs from the current posterior, BEAST halts and reports the difference. To find the bug that caused this dreaded problem, it is handy to find out which operator caused the last state change, and thus which `CalculationNodes` might not have updated themselves properly. Have a look at the `MCMC doloop` method and the debugging code inside for further details.

14.6.2   Trees with traits

A common problem is to associate properties to various components of a tree, for example, rate categories in the relaxed clock model. The easiest way to associate a trait to the nodes (or branches) of a tree is to define a parameter with dimension equal to the number of nodes (or branches). Nodes in the tree with n leaf nodes, so $2n - 1$ nodes in total, are numbered as follows.

- Leaf nodes are numbered $0, \ldots, n - 1$.
- Internal nodes are numbered $n, \ldots, 2n - 1$.
- Root node is not treated as a special internal node, so no number is guaranteed.

This way entry k in the parameter can be associated with the node numbered k. Some extra care needs to be taken when the trait is associated with branches instead of nodes, since the root node is not automatically numbered $2n - 1$. The `TreeWithMetaData Logger` is useful for logging traits on trees.

14.7   Known ways to get into trouble

An `Input` is not declared public

If `Inputs` are not public, they cannot get values assigned by, for instance, the `XMLParser`. The parser will fail to recognise the input and will suggest to assign the value to one of the known public inputs if the input is specified in the XML.

Type of input is a template class (other than `List`)

Thanks to limitations of Java introspection, `Inputs` should be of a type that is concrete, and apart from `List` no template class should be used since it is not possible to determine the type of input automatically otherwise. If you really need an input that is a template class, there are Input constructors where you can provide the class as an extra argument.

Store/restore do not call `super.store()`/`super.restore()`

Obviously, not calling store/restore on superclasses may result in unexpected behaviour. The base implementation in `CalculationNode` sets a flag indicating that the current state is clean, and not setting the flag may lead to inefficient calculation of the posterior.

Derive from BEASTObject **instead of** CalculationNode
Every BEAST-object in a model in between a StateNode that is part of the State and the posterior input of MCMC must be a CalculationNode for efficient model updating to work properly. The framework will point out when this mistake is made.

Using java.util.Random **instead of** beast.util.Randomizer
The seed passed on from the command line is used to initialise Randomizer, not Random. So, rerunning the same XML with the same seed will result in exactly the same chain, which makes debugging easier when using Randomizer.

Using a reserved input name
It is often tempting to use name as the name for an input, but this is reserved for special use by the XML parser. The other reserved names are id, idref and spec. Running the XMLElementNameTest test will warn for such a mistake.

Using an input name already defined in superclass
Input names should be unique, but it can happen that by mistake an input is given a name that is already defined in a superclass. Again, running the XMLElementNameTest test will warn for such a mistake.

Unit tests are missing
Testing is a tedious but necessary part of development. Packages can be set up to be tested every time code is checked in through continuous automation testing software like Hudson. In fact, at http://hudson.cs.auckland.ac.nz/ you can see the state of the latest development code and a number of packages. You can configure your build scripts so that generic tests like the XMLElementNameTest, ExampleXmlParsingTest and DocumentationTest are run on your code as well.

Improper description or input tip text provided in BEAST-object
Obviously.

14.8   Exercise

Write a clock model that, like the uncorrelated relaxed clock, selects a rate for a branch, but where the number of different categories is limited to a given upper bound. Implement it as a lean CalculationNode. Optimise the class by implementing it as a fat CalculationNode.

15 Putting it all together

15.1 Introduction

BEAST 1 does a fantastic job in performing a wide range of phylogenetic analyses. The success of BEAST 1 has meant that many researchers started using it and that demand has fuelled tremendous growth in its source code base. An unintended side-effect was that it made it hard for newcomers to learn the code. Since a lot of the work done with BEAST is at the cutting edge of phylogenetic research, parts of the code base are used to explore new ideas. Since not all ideas work out as expected, some experimental code is abandoned. However, only if you know what to look for is it clear which classes are experimental and which are production code. A partial solution to these problems in BEAST 2 is the *package*. A package is a library based on BEAST 2 that can be installed separately from the BEAST 2 core libraries. This way, the core BEAST 2 library remains small and all its classes are production code. This makes it easier for new developers and PhD students in phylogenetic research to learn BEAST 2. It also makes development work less cumbersome, so that BEAST 2 acts more like a platform for Bayesian phylogenetics, rather than a single monolithic code base. The platform is intended to be quite stable and individual researchers can develop and advertise packages independently of the BEAST 2 release cycle. This provides a cleaner mechanism for providing correct attribution of research, as individual packages can be published separately. Furthermore, it separates out experimental code from the core, which makes it easier to determine which classes are relevant and which are not.

Users can install packages effortlessly through the package manager in BEAUti. Developers can check out code from the package repository. Some packages already available are:

- SNAPP for performing multispecies coalescent analysis for SNP and AFLP data;
- BDSKY contains a birth–death skyline tree prior;
- subst-BMA for Bayesian model averaging over non-contiguous partitions and substitution models;
- RB contains a reversible-jump substitution model and auto-partition functionality;
- BEASTlabs has a range of utilities such as multi-chain MCMC, some experimental methods of inference, a number of experimental likelihood cores;

- MASTER is a framework for simulation studies;
- MultiTypeTree contains classes for using the structured coalescent;
- BEASTShell for scripting BEAST, facilitating *ad hoc* exploration, workflow for simulation studies and creating JUnit tests;
- BEAST-classic has a tool for porting classes from BEAST 1 and contains classes that facilitate ancestral reconstruction and continuous phylogeography.

A number of other packages are in development. To get an overview of the most recently available packages, keep an eye on the BEAST 2 wiki.

In the next section we will have a look at what constitutes a package and how to let the world know the package is ready for use. Making a package available via a GUI like BEAUti can increase its popularity. In Section 15.3 we will have a look at BEAUti and its templates. We give a detailed example to illustrate the various aspects of package development in Section 15.4.

15.2   What is a package?

A package consists of the following components:

- A jar file that contains the class files of BEAST-objects, all supporting code and potentially some classes for supporting BEAUti. Other libraries used for developing the package can be added separately.
- A jar file with the source code. BEAST 2 is licensed under LGPL, since the BEAST team are strong advocates for open source software. All derived work should have its source code made available.
- Example XML files illustrating typical usage of the BEAST-object, similar to the example files distributed with BEAST 1.
- Documentation describing the purpose of the package and perhaps containing articles that can serve as a reference for the package.
- A BEAUti 2 template can be added so that any BEAST-objects in the package are directly available for usage in a GUI.
- A file named `version.xml` which contains the name and version of the package, and the names and versions of any other packages it depends on. For instance, the `version.xml` file for a package which depends only on the current release of BEAST would have the following simple structure:

```
1  <addon name='myPackage' version='1.0.0'>
2    <depends on='beast2' atleast='2.2.0'/>
3  </addon>
4
```

The package consists of a zip-archive containing all of the above. To make the package available for other BEAST users, the zip-file should be downloadable from a public URL. To install a package, download it and unzip it in the `beast2` directory, which is a different location depending on the operating system.[1] This is best done through BEAUti, but can also be done from the command line using the `packagemanager` program. BEAST and BEAUti will automatically check out these directories and load any of the available jar files.

BEAST expects packages to follow the following directory structure:

`myPackage.src.jar` source files
`examples/` XML examples
`lib/` libraries used (if any)
`doc/` documentation
`templates/` BEAUti templates (optional)
`version.xml` Package metadata file

The natural order in which to develop a package is to develop one or more BEAST-objects, test (see Section 15.4) and document these BEAST-objects, develop example XML files and then develop BEAUti support. To start package development, you need to check out the BEAST 2 code and set up a new project that has a dependency on BEAST 2. Details for setting up a package in IDEs like Eclipse or Intellij can be found on the BEAST 2 wiki. Directions on writing BEAST-objects were already discussed in Chapter 14.

15.3 BEAUti

To make a package popular it is important to have GUI support, since many users are not keen to edit raw BEAST XML files. BEAUti is a panel-based GUI for manipulating BEAST models and reading and writing XML files. Unfortunately, a lot of concepts are involved in GUI development, as well as understanding BEAST models. Consequently, this section is rather dense, so be prepared for a steep learning curve. This sections aims at introducing the basic concepts, but once you read it you will probably want to study the templates that come with the BEAST distribution as well before writing your own.

The easiest way to make a BEAST-object available is to write a BEAUti template, which can define new BEAUti panels and determines which sub-models go in which BEAUti panel. A BEAUti template is stored in an XML file in BEAST XML format. There are two types of templates: main templates and sub-templates. Main templates define a complete analysis, while sub-templates define a sub-model, for example a substitution model, which can be used in main templates like the Standard or *BEAST template. Main templates are specified using a `BeautiConfig` object, while sub-templates are specified through `BeautiSubTemplate` objects.

---

[1] The package directory is `$user.home/BEAST` for Windows, `$user.home/Library/Application Support/BEAST` for Mac OS X, `$user.home/.beast2` for Linux, where `$user.home` is the user's home directory for the relevant OS.

15.3.1 BEAUti components

The important objects in a template are `BeautiPanelConfig` and `Beauti Subtemplate`. `BeautiPanelConfig` objects define the part of a model shown in a panel. A panel can show one BEAST-object or a list of BEAST-objects. As you would expect, BEAUti uses the fact that everything in BEAST is a BEAST-object connected to other BEAST-objects through their inputs. In fact, one way to look at BEAUti is as an application that has the task of setting the values of inputs of BEAST-objects. There are three types of inputs: primitive, BEAST-object and list of BEAST-objects. For primitive inputs such as integer and string values, this is just a matter of allowing the user to enter a value. For BEAST-object inputs there are possibly other BEAST-objects that can take their place. The task for BEAUti is connecting and disconnecting BEAST-objects with the proper inputs. For list inputs, BEAUti is allowed to connect more than one BEAST-object to such input.

A `BeautiSubTemplate` specifies a sub-graph and a list of rules on how to connect and disconnect the sub-graph with other components in the graph. These rules are contained in `BeautiConnector` BEAST-objects and there are often many of these rules. A `BeautiConfig` BEAST-object has one special `BeautiSubTemplate`, the *partition-template*, which is instantiated whenever a new partition is created.

Note that some panels are only defined in the context of a partition (e.g. a clock model panel) while others (e.g. the MCMC panel) are not. The `BeautiPanelConfig` BEAST-object needs to know if this is the case and which part of the partition (clock, site or tree model) forms its context.

The description here is just a high-level overview of the way BEAUti works and is configured. For a detailed description of the inputs of various BEAUti BEAST-objects, we refer to the description in the code.

15.3.2 Input editors for BEAUti

BEAUti essentially is a program for manipulating inputs of BEAST-objects. The `InputEditor` class is the base class for all input editors and a number of standard input editors is already implemented. There are input editors for primitives (String, Integer, Boolean, Real) and a generic `BEASTObjectInputEditor`. The latter just lists all inputs of a BEAST-object. There are also more specialised input editors, such as the `TipDatesInputEditor` for editing date information for taxa. Every input editor lists the types of inputs it can handle, which BEAUti uses to determine which input editor to use.

The `BEASTObjectInputEditor` shows the id of the BEAST-object, say X, in a combo-box. Since all but one BEAST-object (typically the MCMC BEAST-object is the exception) is an input to another BEAST-object, BEAUti can determine the type of the input and suggest a replacement of BEAST-object X. BEAUti looks at the list of sub-templates, and the type of the sub-template tells BEAUti whether the sub-template is compatible with the parent input. If so, it is listed in the combo-box as a replacement. It is also possible to expand X and show inputs of BEAST-object X by settings in the template. When the inputs of X are expanded, for every input of X a suitable input editor is found and added to the panel. It is also possible to suppress some of the inputs of X in order to keep the GUI looking clean or hide some of the more obscure options.

There are a number of ways to define the behaviour of an input editor, such as whether to show all inputs of a BEAST-object, and whether to suppress some of the inputs. For list input editors, buttons may be shown to add, remove or edit items from a list. For example, for the list operators it makes sense to allow editing, but not adding.

To write an input editor, derive a class from `InputEditor` or from some class that already derives from `InputEditor`. In particular, if the input editor you want to write can handle list inputs, then derive from `ListInputEditor`.

The important methods to implement are `type` or `types` and `init`. The `type` method tells BEAUti to use this input editor for the particular input class (use `types` if multiple classes are supported). The `init` method should add components containing all the user-interface components for manipulating the input. This is where custom-made code can be inserted to create the desired user-interface for an input. A complex example is the tip-dates input editor, which can be used to edit the traits-input of a tree. Though the tip dates are simply encoded as comma-separated strings with `name=value` pairs, it is much more desirable to manipulate these dates in a table, which is what the tip-dates input editor does.

Input editors are discovered by BEAUti through Java introspection, so they can be part of any jar file in any package. However, they are only expected to be in the `beast.app` package and won't be picked up from other packages.

For more details, have a look at existing input editors in BEAST. It is often easy to build input editors out of existing ones, which can save quite a bit of boilerplate code.

15.3.3   BEAUti template file format

A good place to become familiar with the format and structure of BEAUti templates is by examining the files in the templates directory installed with BEAST. `Standard.xml` is a main template for the most common kind of analysis. It relies on a number of merge points so sub-templates can specify substitution models, clock models and parametric distributions through packages that can be merged into a main template (see end of this section).

The larger part of the standard template consists of a `BeautiConfig` object specification and there is a short XML fragment at the end specifying a MCMC object with a bare-bones analysis. As soon as an alignment is imported in BEAUti, the partition-template in the `BeautiConfig` object will be instantiated and the MCMC graph will be connected with the appropriate tree-likelihoods, priors, operators, loggers and initialisers. When an alignment is imported a name for the alignment is deduced and wherever the phrase (n) occurs in a sub-template it will be replaced by the name of the alignment.

Box 15.1 describes the Yule tree prior, which is a typical template with a parameter for the birth rate and prior on this birth rate. Line 1 defines the sub-template for partition (n).

Box 15.1 BEAUti sub-template for a Yule model.

```
1 <subtemplate id='YuleModel' spec='BeautiSubTemplate' class='beast.
    evolution.speciation.YuleModel' mainid='YuleModel.t:$(n)'>
2 <![CDATA[
3     <object spec='YuleModel' id="YuleModel.t:$(n)" tree='@Tree.t:$
    (n)'>
4         <parameter name='birthDiffRate' id="birthRate.t:$(n)"
    value='1.0'/>
5     </object>
6     <prior id='YuleBirthRatePrior.t:$(n)' spec='Prior' x='
    @birthRate.t:$(n)'><distr spec="beast.math.distributions.
    Uniform" lower='0' upper='1000'/></prior>
7     <scale id='YuleBirthRateScaler.t:$(n)' spec='ScaleOperator'
    scaleFactor="0.75" weight="3" parameter="@birthRate.t:$(n)"/>
8 ]]>
9     <connect spec='BeautiConnector' srcID='YuleModel.t:$(n)'
    targetID='prior' inputName='distribution' if='inposterior
    (YuleModel.t:$(n)) and Tree.t:$(n)/estimate=true'/>
10    <connect spec='BeautiConnector' srcID='birthRate.t:$(n)'
    targetID='state' inputName='stateNode' if='inposterior
    (YuleModel.t:$(n)) and birthRate.t:$(n)/estimate=true'/>
11    <connect spec='BeautiConnector' srcID='YuleBirthRatePrior.t:$
    (n)' targetID='prior' inputName='distribution'  if='inposterior
    (YuleModel.t:$(n)) and birthRate.t:$(n)/estimate=true'/>
12    <connect spec='BeautiConnector' srcID='YuleBirthRateScaler.t:$
    (n)' targetID='mcmc' inputName='operator'        if='inposterior
    (YuleModel.t:$(n)) and birthRate.t:$(n)/estimate=true'/>
13    <connect spec='BeautiConnector' srcID='YuleModel.t:$(n)'
    targetID='tracelog' inputName='log' if='inposterior(YuleModel.
    t:$(n)) and Tree.t:$(n)/estimate=true'/>
14    <connect spec='BeautiConnector' srcID='birthRate.t:$(n)'
    targetID='tracelog' inputName='log' if='inposterior(YuleModel.
    t:$(n)) and birthRate.t:$(n)/estimate=true'/>
15 </subtemplate>
```

A sub-template has an id that is used in combo-boxes in BEAUti to identify the sub-template. The `class` attribute indicates that the sub-template produces a BEAST-object of this type. One task of BEAUti is to create new BEAST-objects, and connect them to an input. The way BEAUti finds potential BEAST-objects to connect to an input is by testing whether the type of an input is compatible with the type of a sub-template. If so, the sub-template is listed in the combo-box for the input in the BEAST-object input editor. The `mainid` attribute identifies the BEAST-object that has the type listed in `class`. Note the `t:$(n)` part of the `mainid` attribute. A sub-template is created in the context of a partition, and wherever there is the string `$(n)` in the sub-template this is replaced by the name of the partition. BEAUti distinguishes three items that form the context of a partition: its site model, clock model and tree. These are identified by having respectively `s:$(n)`, `c:$(n)` and `t:$(n)` in their id. In the Yule prior template, the prior, parameter and operators are all in the context of a tree, so these all have `t:$(n)` in their ids.

The CDATA section from lines 2 to 8 contains the sub-graph created when the template is activated. For a Yule prior on the tree (line 3), a prior on the birth rate (line 6) and a scale operator on the birth rate (line 7) are created. Lines 9 to 14 specify the connections that need to be made through `BeautiConnectors`. A connector specifies a BEAST-object to connect from (through the `srcID` attribute), a BEAST-object to connect to (through the `targetID` attribue) and an `inputName` specifies the name of the input in the target BEAST-object to connect with. The connector is only activated when some conditions are met. If the condition is not met, BEAUti will attempt to disconnect the link (if it exists). The conditions are separated by the 'and' keyword in the `if` attribute. The conditions are mostly of the form `inposterior(YuleModel.t:$(n))`, which tests whether the BEAST-object with id `YuleModel.t:$(n)` is a predecessor of a BEAST-object with id `posterior` in the model. Further, there are conditions of the form `Tree.t:$(n)/ estimate=true`, used to test whether an input value of a BEAST-object has a certain value. This is mostly relevant to test whether `StateNodes` are estimated or not, since if they are not estimated no operator should be defined on it, and logging is not very useful.

Line 9 connects the prior to the BEAST-object with id 'prior'. This refers to a compound distribution inside the MCMC, and the Yule prior is added to the input with the name 'distribution'. The birth rate parameter is added to the state (line 10), the prior on the birth rate is added to the prior (line 11), the scale operator is connected to the MCMC operators input (line 12) and the Yule prior and birth rate are added to the trace log (lines 13 and 14). Note that these connections are only executed if the condition specified in the `if` input is true, otherwise the connection is attempted to be disconnected.

Connectors are tested in order of appearance. It is always a good idea to make the first connector the one connecting the main BEAST-object in the sub-template, since if this main BEAST-object is disconnected, most of the others should be disconnected as well. For this tree prior, the tree's `estimate` flag can become false when the tree for the partition is linked.

Instead of defining all sub-templates explicitly for the `BeautiConfig` BEAST-object, a merge-point can be defined. Before processing a template, all merge-points are replaced by XML fragments in sub-templates. For example, the main template can contain `<mergepoint id='parametricDistributions'/>` inside a `BeautiConfig` BEAST-object and a sub-template can contain an XML fragment like this:

```
1  <mergewith point='substModelTemplates'>
2    <subtemplate id='JC69' class='beast.evolution.substitutionmodel.
       JukesCantor' mainid='JC69.s:$(n)'>
3  <![CDATA[
4      <distr spec='JukesCantor' id='JC69.s:$(n)'/>
5  ]]>
6    </subtemplate>
7  </mergewith>
```

Line 1 has a `mergewith` element that refers to merge-point with id `substModel Templates` defined in the main template. Lines 2 to 6 define a sub-template that will be inserted in the XML of the main template. The sub-template needs an id, a class specifying the class of the input it can be connected to and the id of the BEAST-object that needs to be connected. The actual BEAST-object that is created when the sub-template is activated is defined inside a CDATA section starting at line 3 with `<![CDATA[` and closing at line 5 with `]]>`. The `BeautiTemplate` has an input called 'value' which contains the XML fragment specifying all BEAST-objects in the sub-graph and everything inside the CDATA section is assigned to that input.

15.4   Variable selection-based substitution model package example

The variable selection-based substitution model (VS model) is a substitution model that jumps between six substitution models using BSVS. The frequencies at the root of the trees are empirically estimated from sequence data. Figure 15.1 shows the parameters involved in the six substitution models. Since BEAST normalises transition probability matrices such that on average one substitution is expected per unit length on a branch, one parameter can be set to 1. The models are selected so that they are nested, that is, every model with i parameters can be expressed in models with j parameters if $j > i$. The popular models F81, HKY85, TN93, TIM and GTR follow this model. In order to finalise the set of models, we chose an extra model, labelled EVS in Figure 15.1, that obeys the nesting constraint. To transition from model i to model $i + 1$ means that the VS model utilised one more variable. If the variable is not used, it is effectively sampled from the prior. In this section we will have a look at what is involved in adding the VS model to BEAST as a package.

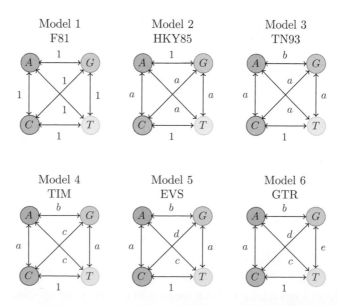

Figure 15.1 The six substitution models and their parameters in the VS substitution model.

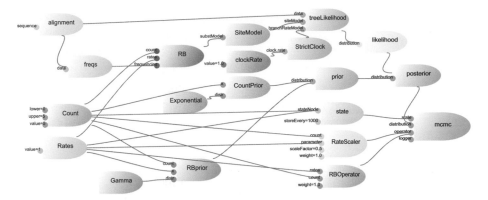

Figure 15.2 The VS model consists of substitution model, two parameters and two operators on these parameters, a prior on the rates and count. The accompanying BEAUti template contains a set of rules to connect its BEAST-objects to a larger model, e.g. the rates to the rate prior, the loggers to the trace-log, etc. For clarity, the tree, sequences and loggers are omitted.

15.4.1   Step 1: Write BEAST-object – develop code

To write a VS model, we go through the following procedure, which is quite typical for developing models;

- Select appropriate `StateNodes` to represent the space we want to sample. The VS model is implemented using an integer parameter as substitution model index, and a six-dimensional real parameter containing the rate parameters.
- Define a set of operators that efficiently sample the state space. There are two operators on these parameters. Firstly, a random walk operator that increases or decreases the dimension of the model. Secondly, a scale operator that selects one of the rate parameters and changes its value.
- Develop classes that ultimately influence the posterior. The VS substitution model derives from `GeneralSubstitutionModel` which takes the count and rate parameters as input. Further, a gamma prior is defined on the rate parameter. Also, we define a prior on the integer parameter so that we can indicate less complex models are preferred over complex models.

Figure 15.2 shows how these elements are linked together.

When services of other classes are required that are package private, the class should be placed in the package of these classes (or otherwise these could not be used). But it is good practice to keep your package in its own name space in order to prevent class name clashes with other packages.

The only new class required for the VS model is a substitution model class that derives from `GeneralSubstitutionModel`. We will have a look at some of the more interesting areas of the implementation. It starts with a proper description:

```
1 @Description("Substitution model for nucleotides that " +
2     "changes where the count input determines the " +
3     "number of parameters used in a hierarchy of models")
```

We need to add an extra input to indicate how many dimensions are in use.

```
1 public Input<IntegerParameter> countInput = new Input<IntegerParameter>
        ("count", "model number used 0 = F81, 1 = HKY, 2 = TN93, 3 = TIM,
        4 = EVS, 5 and higher GTR (default 0)", Validate.REQUIRED);
```

The `initAndValidate()` method initialises some parameters, which is fairly straightforward. Then, we override `setupRelativeRates` where the rates are initialised in one large case-statement that switches based on the count input.

```
1  @Override
2  protected void setupRelativeRates() {
3      switch (count.getValue()) {
4      case 0: // JC96
5          Arrays.fill(relativeRates, 1.0);
6          break;
7      case 1: // HKY
8          relativeRates[0] = m_rate.getArrayValue(0); // A->C
9          relativeRates[1] = 1; // A->G
10         relativeRates[2] = m_rate.getArrayValue(0); // A->T
11     // see source code for more...
12     ...
13     }
14 }
```

Since the substitution model is intended for nucleotide data only, we need to add a method indicating that other data-types are not supported. To this end, we override the `canHandleDataType` method.

```
1  @Override
2  public boolean canHandleDataType(DataType dataType) throws Exception {
3      if (dataType instanceof Nucleotide) {
4          return true;
5      }
6      throw new Exception("Can only handle nucleotide data");
7  }
```

The superclass takes care of everything else, including storing, restoring and setting a flag for recalculating items. A small efficiency gain could be achieved by letting the `requiresRecalculation` method test whether any of the relevant rates changed, which is left as an exercise to the reader.

15.4.2   Step 2: Test the model

Unfortunately, models require testing. There are various model testing strategies:

- unit tests to ensure the code is correct;
- prior sampling to ensure operators work correctly and priors do not interfere with each other;
- integration testing where we sample data from the prior, then run an MCMC analysis with the model to ensure the model can indeed be recovered.

For unit testing of BEAST-objects, JUnit tests are the *de facto* standard for Java programs. There are some general tests in BEAST you might want to use for checking that a BEAST-object is properly written. There is the `DocumentationTest` for testing whether description annotations and input tip texts are up to scratch and `XMLElementNameTest` to check that input names are not accidentally chosen to be reserved names, since using reserved names can lead to unexpected and hard-to-find bugs. For some reason, the name 'name' is a desirable name for an input, which unfortunately is already reserved. The `ExampleXmlParsingTest` tests whether example XML parses and runs for a limited number of samples. We found it useful to set up an automatic testing system, which runs JUnit tests every time code is checked in, and run integration tests once a day.

In the case of the VS substitution model, we write tests to make sure the model behaves the same as some known substitution models behave. This gives us a test for F81, HKY, TN93, etc. Next, we create an example XML file, which goes in the examples directory that goes with the package. This example file can work as a blueprint for the BEAUti template to a large extent.

One important step that is regularly overlooked in testing a new model is to see how the model behaves when sampling from the prior. The reason to test this is that often subtle errors can occur that are not easily noticed when sampling from a full model. For example, in the VS model we want to make sure that every model is equally often sampled when there is a uniform prior on the count. It is easy to come up with an operator that randomly increments or decrements the count. But in the case the count is zero, no decrement is possible. If the operator would just increase the counter in this situation the distribution over models would be lower for F81 than for HKY85, TN93, TIM, EVS and GTR, which all would be equally likely. The random walk operator fails when the count parameter is zero and it tries to decrease, which ensures all counts are equally likely.

Another reason to sample from the prior is that priors may interact in unexpected and perhaps undesirable ways. Sampling from the prior makes it clear whether there is any such interaction. For the VS model, we make sure that the count parameter is uniformly distributed from 0 to 5, and that the rate parameters all have the same mean as the prior over the rates, which is 1 in this case.

Integration testing is best done by using a simulator that generates data you want to analyse and determine how well the model used for generating the data is recovered by the model based on the new BEAST-objects. There is some support in BEAST 2 for generating alignment data through the `beast.app.SequenceSimulator` class, which provides methods to generate sequence alignments for a fixed tree with various substitution models, site models and branch rate models. To run it you need an XML specification which uses `SequenceSimulator` in the run element. For more advanced simulation studies the MASTER package may be of interest.

For the VS model, we want to sample sequence data from the model, and use this data to recover the model in an MCMC analysis. For each of the six substitution models, we can generate a random substitution model with rates sampled from the prior used on the rates. Note that we sample from the prior over the parameters to ensure the priors used are sensible. Then we use the sequence simulator to sample an alignment for each of the substitution models with the sampled rates. Finally, we run an analysis with the VS model and check how close the substitution model that is estimated is to the one used to generate the data. Ideally, when sampling from the VS model with *x* parameters, we obtain an estimate of *x* parameters when analysing the data. The proportion of time the number of parameters is indeed estmated as *x* is a measure of how well the model performs.

There is quite a complex workflow involved in setting up such tests, and much of this is repetitive since we want to run through the process multiple times. You might find the BEASTShell package useful for this, as it was designed to help with testing BEAST models.

15.4.3   Step 3: Write BEAUti template

Before writing the BEAUti template, you need to understand some of the basics of BEAUti as outlined in Section 15.3. The VS model is a substitution model; this model should be made available as a choice in the site model panel whenever there is a nucleotide partition. The complete code is available in the VSS package, but here we highlight some of it. Since a BEAUti template is in BEAST XML format, we start with a BEAST 2 header followed by the sub-template definition:

```
 8 <subtemplate id='VS' class='beast.evolution.substitutionmodel.VS'
      mainid='VS.s:$(n)'
 9         suppressInputs='beast.evolution.substitutionmodel.VS.
      eigenSystem,...'>
```

Some inputs, like `eigenSystem`, should not show in BEAUti, which is marked in this declaration. At its heart, the template consists of a BEAST XML fragment specifying the model, its priors and operators, wrapped in a CDATA section.

```
10 <![CDATA[
11     <substModel spec='VS' id='VS.s:$(n)'>
12         <count spec='parameter.IntegerParameter' id='VScount.s:$(n)'
      value='5' lower='0' upper='5'/>
13         <rates spec='parameter.RealParameter' id='VSrates.s:$(n)'
      value='1' dimension='5' lower='0.01' upper='100.0'/>
14         <frequencies id='freqs.s:$(n)' spec='Frequencies'>
15             <data idref='$(n)'/>
16         </frequencies>
17     </substModel>
18
19     <distribution id='VSprior.s:$(n)' spec='beast.math.distributions.
      Prior' x='@VSrates.s:$(n)'>
20         <distr spec="beast.math.distributions.Gamma" alpha='0.2'
      beta='5.0'/>
21     </distribution>
22
23     <operator id='VSOperator.s:$(n)' spec='IntRandomWalkOperator'
      weight="9" windowSize='1' parameter='@VScount.s:$(n)'/>
24     <operator id='VSratescaler.s:$(n)' spec='ScaleOperator'
      scaleFactor="0.5" weight="9" parameter="@VSrates.s:$(n)"/>
25 ]]>
```

The remainder of the sub-template consists of BEAUti-connector rules, for example for connecting the VScount-parameter to the state and the trace logger.

```
26 <connect srcID='VScount.s:$(n)'        targetID='state'  inputName=
     'stateNode'  if='inlikelihood(VScount.s:$(n))'/>
27 <connect srcID='VSrates.s:$(n)'        targetID='state'  inputName=
     'stateNode'  if='inlikelihood(VSrates.s:$(n))'/>
28 ...
```

15.4.4 Step 4: Update build script to create package

Since a package is a zip-file that contains some bits and pieces (Section 15.2), it is probably easiest to have a look at a build file for an existing package and copy the relevant parts to your own build file.

15.4.5 Step 5: Publish package by link on website

The BEAST website has further instructions on how to publish the package so that the package manager (accessible in BEAUti) can pick it up and install it on user computers or computer clusters.

The page http://beast2.org/wiki/index.php/VSS_package has instructions on how to install the complete package so you can inspect the code and BEAUti template.

15.5 Exercise

Write an exciting new package for BEAST 2 and release it to the public!

References

Akaike, H (1974). 'A new look at the statistical model identification'. In: *IEEE Transactions on Automatic Control* 19.6, pp. 716–723 (page 18).

Aldous, D (2001). 'Stochastic models and descriptive statistics for phylogenetic trees, from Yule to today'. In: *Statistical Science* 16, pp. 23–34 (pages 31, 39).

Alekseyenko, AV, CJ Lee and MA Suchard (2008). 'Wagner and Dollo: a stochastic duet by composing two parsimonious solos'. In: *Systematic Biology* 57.5, pp. 772–784 (page 115).

Allen, LJ (2003). *An introduction to stochastic processes with applications to biology*. Upper Saddle River, NJ: Pearson Education (page 12).

Amenta, N and J Klingner (2002). 'Case study: visualizing sets of evolutionary trees'. In: *INFOVIS 2002. IEEE Symposium on Information Visualization*, pp. 71–74 (page 159).

Anderson, RM and RM May (1991). *Infectious diseases of humans: dynamics and control*. Oxford: Oxford University Press (page 34).

Arunapuram, P, I Edvardsson, M Golden, et al. (2013). 'StatAlign 2.0: combining statistical alignment with RNA secondary structure prediction'. In: *Bioinformatics* 29.5, pp. 654–655 (page 9).

Atarhouch, T, L Rüber, EG Gonzalez, et al. (2006). 'Signature of an early genetic bottleneck in a population of Moroccan sardines (*Sardina pilchardus*)'. In: *Molecular Phylogenetics and Evolution* 39.2, pp. 373–383 (page 133).

Avise, J (2000). *Phylogeography: the history and formation of species*. Cambridge, MA: Harvard University Press (page 68).

Ayres, D, A Darling, D Zwickl, et al. (2012). 'BEAGLE: a common application programming interface and high-performance computing library for statistical phylogenetics'. In: *Systematic Biology* 61, pp. 170–173 (page 114).

Baele, G, P Lemey, T Bedford, et al. (2012). 'Improving the accuracy of demographic and molecular clock model comparison while accommodating phylogenetic uncertainty'. In: *Molecular Biology and Evolution* 29.9, pp. 2157–2167 (pages 18, 19, 136, 137).

Baele, G, WLS Li, AJ Drummond, MA Suchard and P Lemey (2013a). 'Accurate model selection of relaxed molecular clocks in Bayesian phylogenetics'. In: *Molecular Biology and Evolution* 30.2, pp. 239–243 (pages 18, 137).

Baele, G, P Lemey and S Vansteelandt (2013b). 'Make the most of your samples: Bayes factor estimators for high-dimensional models of sequence evolution'. In: *BMC Bioinformatics* 14.1, p. 85 (page 19).

Bahl, J, MC Lau, GJ Smith, et al. (2011). 'Ancient origins determine global biogeography of hot and cold desert cyanobacteria'. In: *Nature Communications* 2, art. 163 (page 96).

Bahlo, M and RC Griffiths (2000). 'Inference from gene trees in a subdivided population'. In: *Theoretical Population Biology* 57.2, pp. 79–95 (page 31).

Barnes, I, P Matheus, B Shapiro, D Jensen and A Cooper (2002). 'Dynamics of Pleistocene population extinctions in Beringian brown bears'. In: *Science* 295, p. 2267 (page 31).

Barrett, M, MJ Donoghue and E Sober (1991). 'Against consensus'. In: *Systematic Zoology* 40.4, pp. 486–493 (page 159).

Barton, NH, F Depaulis and AM Etheridge (2002). 'Neutral evolution in spatially continuous populations'. In: *Theoretical Population Biology* 61.1, pp. 31–48 (page 74).

Barton, NH, J Kelleher and AM Etheridge (2010a). 'A new model for extinction and recolonization in two dimensions: quantifying phylogeography'. In: *Evolution* 64.9, pp. 2701–2715 (page 74).

Barton, NH, AM Etheridge and A Veber (2010b). 'A new model for evolution in a spatial continuum'. In: *Electronic Journal of Probability* 15, pp. 162–216 (page 74).

Beaumont, MA, R Nielsen, C Robert, et al. (2010). 'In defence of model-based inference in phylogeography'. In: *Molecular Ecology* 19.3, pp. 436–446 (page 69).

Bedford, T, S Cobey, P Beerli and M Pascual (2010). 'Global migration dynamics underlie evolution and persistence of human influenza A (H3N2)'. In: *PLoS Pathogens* 6.5, e1000918 (page 71).

Beerli, P (2004). 'Effect of unsampled populations on the estimation of population sizes and migration rates between sampled populations'. In: *Molecular Ecology* 13.4, pp. 827–836 (page 72).

Beerli, P (2006). 'Comparison of Bayesian and maximum-likelihood inference of population genetic parameters.' In: *Bioinformatics* 22.3, pp. 341–345 (page 14).

Beerli, P and J Felsenstein (1999). 'Maximum-likelihood estimation of migration rates and effective population numbers in two populations using a coalescent approach'. In: *Genetics* 152.2, pp. 763–773 (pages 6, 31).

Beerli, P and J Felsenstein (2001). 'Maximum likelihood estimation of a migration matrix and effective population sizes in *n* subpopulations by using a coalescent approach.' In: *Proceedings of the National Academy of Sciences USA* 98.8, pp. 4563–4568 (pages 6, 31, 69, 71, 72).

Belfiore, N, L Liu and C Moritz (2008). 'Multilocus phylogenetics of a rapid radiation in the genus *Thomomys* (Rodentia: Geomyidae)'. In: *Systematic Biology* 57.2, pp. 294–310 (page 120).

Benner, P, M Bačák and PY Bourguignon (2014). 'Point estimates in phylogenetic reconstructions'. In: *Bioinformatics* 30.17, pp. i534–i540 (page 154).

Berger, JO (2006). 'The case for objective Bayesian analysis'. In: *Bayesian Analysis* 1.3, pp. 385–402 (page 20).

Berger, JO and JM Bernardo (1992). 'On the development of reference priors'. In: *Bayesian statistics*. Ed. by JM Bernardo, JO Berger, AP Dawid and AFM Smith. Vol. 4. Oxford: Oxford University Press, pp. 35–60 (pages 14, 127).

Biek, R, AJ Drummond and M Poss (2006). 'A virus reveals population structure and recent demographic history of its carnivore host'. In: *Science* 311, pp. 538–541 (pages 73, 133).

Biek, R, JC Henderson, LA Waller, CE Rupprecht and LA Real (2007). 'A high-resolution genetic signature of demographic and spatial expansion in epizootic rabies virus'. In: *Proceedings of the National Academy of Sciences USA* 104.19, pp. 7993–7998 (page 73).

Bielejec, F, A Rambaut, MA Suchard and P Lemey (2011). 'SPREAD: spatial phylogenetic reconstruction of evolutionary dynamics'. In: *Bioinformatics* 27.20, pp. 2910–2912 (page 135).

Bielejec, F, P Lemey, G Baele, A Rambaut and MA Suchard (2014). 'Inferring heterogeneous evolutionary processes through time: from sequence substitution to phylogeography'. In: *Systematic Biology* 63.4, pp. 493–504 (page 113).

Billera, L, S Holmes and K Vogtmann (2001). 'Geometry of the space of phylogenetic trees'. In: *Advances in Applied Mathematics* 27, pp. 733–767 (page 154).

Bloomquist, EW and MA Suchard (2010). 'Unifying vertical and nonvertical evolution: a stochastic ARG-based framework'. In: *Systematic Biology* 59.1, pp. 27–41 (pages 6, 10).

Bofkin, L and N Goldman (2007). 'Variation in evolutionary processes at different codon positions'. In: *Molecular Biology and Evolution* 24.2, pp. 513–521 (pages 100, 104, 146).

Bolstad, WM (2011). *Understanding computational Bayesian statistics*. Vol. 644. Hoboken, NJ: Wiley (page 10).

Boni, MF, D Posada and MW Feldman (2007). 'An exact nonparametric method for inferring mosaic structure in sequence triplets'. In: *Genetics* 176.2, pp. 1035–1047 (page 97).

Bouckaert, RR (2010). 'DensiTree: making sense of sets of phylogenetic trees'. In: *Bioinformatics* 26.10, pp. 1372–1373 (pages 156, 163).

Bouckaert, RR and D Bryant (2012). 'A rough guide to SNAPP'. In: *BEAST 2 wiki* (pages x, 124).

Bouckaert, RR, P Lemey, M Dunn, et al. (2012). 'Mapping the origins and expansion of the Indo-European language family'. In: *Science* 337, pp. 957–960 (pages 57, 135).

Bouckaert, RR, M Alvarado-Mora and J Rebello Pinho (2013). 'Evolutionary rates and HBV: issues of rate estimation with Bayesian molecular methods'. In: *Antiviral Therapy* 18, pp. 497–503 (pages 57, 100, 145).

Bouckaert, RR, J Heled, D Kühnert, et al. (2014). 'BEAST 2: a software platform for Bayesian evolutionary analysis'. In: *PLoS Computational Biology* 10.4, e1003537 (page 16).

Bradley, RK, A Roberts, M Smoot, et al. (2009). 'Fast statistical alignment'. In: *PLoS Computational Biology* 5.5, e1000392 (page 9).

Brooks, S and A Gelman (1998). 'Assessing convergence of Markov chain Monte Carlo algorithms'. In: *Journal of Computational and Graphical Statistics* 7, pp. 434–455 (page 140).

Brooks, S, A Gelman, GL Jones and XL Meng (2010). *Handbook of Markov chain Monte Carlo*. Boca Raton, FL: Chapman & Hall/CRC. (page 10).

Brown, JM (2014). 'Predictive approaches to assessing the fit of evolutionary models'. In: *Systematic Biology* 63.3, pp. 289–292 (page 18).

Brown, RP and Z Yang (2011). 'Rate variation and estimation of divergence times using strict and relaxed clocks'. In: *BMC Evolutionary Biology* 11, art. 271 (pages 105, 106, 144, 146).

Bryant, D (2003). 'A classification of consensus methods for phylogenetics'. In: *BioConsensus (Piscataway, NJ, 2000/2001)*. Providence, RI: AMS, pp. 163–184 (page 159).

Bryant, D, RR Bouckaert, J Felsenstein, N Rosenberg, and A RoyChoudhury (2012). 'Inferring species trees directly from biallelic genetic markers: bypassing gene trees in a full coalescent analysis'. In: *Molecular Biology and Evolution* 29.8, pp. 1917–1932 (pages 7, 115, 119, 123).

Bunce, M, TH Worthy, T Ford, et al. (2003). 'Extreme reversed sexual size dimorphism in the extinct New Zealand moa *Dinornis*.' In: *Nature* 425, pp. 172–175 (page 129).

Burnham, K and D Anderson (2002). *Model selection and multimodel inference: a practical information-theoretic approach*. New York: Springer Verlag (page 32).

Camargo, A, LJ Avila, M Morando and JW Sites (2012). 'Accuracy and precision of species trees: effects of locus, individual, and base pair sampling on inference of species trees in lizards of the *Liolaemus darwinii* group (Squamata, Liolaemidae)'. In: *Systematic Biology* 61.2, pp. 272–288 (page 120).

Campos, PF, E Willerslev, A Sher, et al. (2010). 'Ancient DNA analyses exclude humans as the driving force behind late Pleistocene musk ox (*Ovibos moschatus*) population dynamics'. In: *Proceedings of the National Academy of Sciences USA* 107.12, pp. 5675–5680 (page 134).

Cavalli-Sforza, L and A Edwards (1967). 'Phylogenetic analysis: models and estimation procedures'. In: *American Journal of Human Genetics* 19, pp. 233–257 (pages 24, 25).

Chaves, JA and TB Smith (2011). 'Evolutionary patterns of diversification in the Andean hummingbird genus *Adelomyia*'. In: *Molecular Phylogenetics and Evolution* 60.2, pp. 207–218 (page 165).

Chung, Y and C Ané (2011). 'Comparing two Bayesian methods for gene tree/species tree reconstruction: simulations with incomplete lineage sorting and horizontal gene transfer'. In: *Systematic Biology* 60.3, pp. 261–275 (page 120).

Coop, G and RC Griffiths (2004). 'Ancestral inference on gene trees under selection'. In: *Theoretical Population Biology* 66.3, pp. 219–232 (pages 69, 76).

Cox, J, J Ingersoll and S Ross (1985). 'A theory of the term structure of interest rates'. In: *Econometrica* 53, pp. 385–407 (page 125).

Currie, TE, SJ Greenhill, R Mace, et al. (2010). 'Is horizontal transmission really a problem for phylogenetic comparative methods? A simulation study using continuous cultural traits'. In: *Philosophical Transactions of the Royal Society B: Biological Sciences* 365, pp. 3903–3912 (page 97).

Debruyne, R, G Chu, CE King, et al. (2008). 'Out of America: ancient DNA evidence for a new world origin of late quaternary woolly mammoths'. In: *Current Biology* 18.17, pp. 1320–1326 (page 133).

Degnan, JH and NA Rosenberg (2006). 'Discordance of species trees with their most likely gene trees'. In: *PLoS Genetics* 2.5, c68 (page 116).

Didelot, X, J Gardy and C Colijn (2014). 'Bayesian inference of infectious disease transmission from whole genome sequence data'. In: *Molecular Biology and Evolution*, msu121 (page 41).

Dincă, V, VA Lukhtanov, G Talavera and R Vila (2011). 'Unexpected layers of cryptic diversity in wood white Leptidea butterflies'. In: *Nature Communications* 2, art. 324 (pages 121, 165).

Drummond, AJ (2002). 'Computational statistical inference for molecular evolution and population genetics'. PhD thesis. University of Auckland (page 106).

Drummond, AJ and A Rambaut (2007). 'BEAST: Bayesian evolutionary analysis by sampling trees'. In: *BMC Evolutionary Biology* 7, art. 214 (pages 14, 74, 79).

Drummond, AJ and AG Rodrigo (2000). 'Reconstructing genealogies of serial samples under the assumption of a molecular clock using serial-sample UPGMA'. In: *Molecular Biology and Evolution* 17.12, pp. 1807–1815 (page 31).

Drummond, AJ and MA Suchard (2008). 'Fully Bayesian tests of neutrality using genealogical summary statistics'. In: *BMC Genetics* 9, art. 68 (page 18).

Drummond, AJ and MA Suchard (2010). 'Bayesian random local clocks, or one rate to rule them all'. In: *BMC Biology* 8.1, art. 114 (pages 63–65, 83, 146).

Drummond, AJ, R Forsberg and AG Rodrigo (2001). 'The inference of stepwise changes in substitution rates using serial sequence samples.' In: *Molecular Biology and Evolution* 18.7, pp. 1365–1371 (page 31).

Drummond, AJ, GK Nicholls, AG Rodrigo and W Solomon (2002). 'Estimating mutation parameters, population history and genealogy simultaneously from temporally spaced sequence data'. In: *Genetics* 161.3, pp. 1307–1320 (pages 6, 14, 31, 33, 34, 66, 74, 109).

Drummond, AJ, OG Pybus, A Rambaut, R Forsberg and AG Rodrigo (2003a). 'Measurably evolving populations'. In: *Trends in Ecology & Evolution* 18, pp. 481–488 (pages 8, 66, 128, 152).

Drummond, AJ, OG Pybus and A Rambaut (2003b). 'Inference of viral evolutionary rates from molecular sequences'. In: *Advances in Parasitology* 54, pp. 331–358 (page 149).

Drummond, AJ, A Rambaut, B Shapiro and OG Pybus (2005). 'Bayesian coalescent inference of past population dynamics from molecular sequences'. In: *Molecular Biology and Evolution* 22.5, pp. 1185–1192 (pages 32, 108, 130, 133).

Drummond, AJ, SYW Ho, MJ Phillips and A Rambaut (2006). 'Relaxed phylogenetics and dating with confidence'. In: *PLoS Biology* 4.5, e88 (pages 62, 63, 73, 144).

Drummond, AJ, MA Suchard, D Xie and A Rambaut (2012). 'Bayesian phylogenetics with BEAUti and the BEAST 1.7'. In: *Molecular Biology and Evolution* 29.8, pp. 1969–1973 (pages 16, 74).

Duffy, S, LA Shackelton and EC Holmes (2008). 'Rates of evolutionary change in viruses: patterns and determinants'. In: *Nature Reviews Genetics* 9.4, pp. 267–276 (pages 20, 152).

Durbin, R, SR Eddy, A Krogh and G Mitchison (1998). *Biological sequence analysis: probabilistic models of proteins and nucleic acids*. Cambridge: Cambridge University Press (page 8).

Edgar, RC (2004a). 'MUSCLE: a multiple sequence alignment method with reduced time and space complexity'. In: *BMC Bioinformatics* 5, p. 113 (page 9).

Edgar, RC (2004b). 'MUSCLE: multiple sequence alignment with high accuracy and high throughput'. In: *Nucleic Acids Research* 32.5, pp. 1792–1797 (page 9).

Edwards, A (1970). 'Estimation of the branch points of a branching diffusion process (with discussion)'. In: *Journal of the Royal Statistical Society, Series B* 32, pp. 155–174 (page 38).

Edwards, A and L Cavalli-Sforza (1965). 'A method for cluster analysis'. In: *Biometrics*, pp. 362–375 (page 6).

Edwards, CTT, EC Holmes, DJ Wilson, et al. (2006). 'Population genetic estimation of the loss of genetic diversity during horizontal transmission of HIV-1'. In: *BMC Evolutionary Biology* 6, art. 28 (page 32).

Etienne, R, B Haegeman, T Stadler, et al. (2012). 'Diversity-dependence brings molecular phylogenies closer to agreement with the fossil record'. In: *Proceedings of the Royal Society B: Biological Sciences,* doi: 10.1098/rspb.2011.1439 (page 39).

Ewing, G and AG Rodrigo (2006a). 'Coalescent-based estimation of population parameters when the number of demes changes over time'. In: *Molecular Biology and Evolution* 23.5, pp. 988–996 (pages 6, 71).

Ewing, G and AG Rodrigo (2006b). 'Estimating population parameters using the structured serial coalescent with Bayesian MCMC inference when some demes are hidden'. In: *Evolutionary Bioinformatics* 2, pp. 227–235 (page 72).

Ewing, G, G Nicholls and AG Rodrigo (2004). 'Using temporally spaced sequences to simultaneously estimate migration rates, mutation rate and population sizes in measurably evolving populations'. In: *Genetics* 168.4, pp. 2407–2420 (pages 6, 14, 69, 71, 72).

Faria, NR, MA Suchard, A Abecasis, et al. (2012). 'Phylodynamics of the HIV-1 CRF02_AG clade in Cameroon'. In: *Infection, Genetics and Evolution* 12.2, pp. 453–460 (page 135).

Fearnhead, P and P Donnelly (2001). 'Estimating recombination rates from population genetic data'. In: *Genetics* 159.3, pp. 1299–1318 (page 31).

Fearnhead, P and C Sherlock (2006). 'An exact Gibbs sampler for the Markov-modulated Poisson process'. In: *Journal of the Royal Statistical Society, Series B* 68.5, pp. 767–784 (page 72).

Felsenstein, J (1981). 'Evolutionary trees from DNA sequences: a maximum likelihood approach'. In: *Journal of Molecular Evolution* 17, pp. 368–376 (pages 6, 49, 53, 55, 69, 70).

Felsenstein, J (1985). 'Phylogenies and the comparative method'. In: *The American Naturalist* 125.1, pp. 1–15 (page 73).

Felsenstein, J (1988). 'Phylogenies from molecular sequences: inference and reliability'. In: *Annual Review of Genetics* 22, pp. 521–565 (page 31).

Felsenstein, J (1992). 'Estimating effective population size from samples of sequences: inefficiency of pairwise and segregating sites as compared to phylogenetic estimates'. In: *Genetical Research* 59, pp. 139–147 (pages 29, 31).

Felsenstein, J (2001). 'The troubled growth of statistical phylogenetics'. In: *Systematic Biology* 50.4, pp. 465–467 (pages 6, 19, 68).

Felsenstein, J (2004). *Inferring phylogenies.* Sunderland, MA: Sinauer Associates (pages 14, 98).

Felsenstein, J (2006). 'Accuracy of coalescent likelihood estimates: do we need more sites, more sequences, or more loci?' In: *Molecular Biology and Evolution* 23.3, pp. 691–700 (pages 6, 131, 151).

Finlay, EK, C Gaillard, S Vahidi, et al. (2007). 'Bayesian inference of population expansions in domestic bovines'. In: *Biology Letters* 3.4, pp. 449–452 (page 133).

Firth, C, A Kitchen, B Shapiro, et al. (2010). 'Using time-structured data to estimate evolutionary rates of double-stranded DNA viruses'. In: *Molecular Biology and Evolution* 27, pp. 2038–2051 (page 152).

Fisher, R (1930). *Genetical theory of natural selection.* Oxford: Clarendon Press (page 28).

FitzJohn, R (2010). 'Quantitative traits and diversification'. In: *Systematic Biology* 59.6, pp. 619–633 (page 39).

FitzJohn, RG, WP Maddison and SP Otto (2009). 'Estimating trait-dependent speciation and extinction rates from incompletely resolved phylogenies'. In: *Systematic Biology* 58.6, pp. 595–611 (page 39).

Ford, CB, PL Lin, MR Chase, et al. (2011). 'Use of whole genome sequencing to estimate the mutation rate of *Mycobacterium tuberculosis* during latent infection'. In: *Nature Genetics* 43.5, pp. 482–486 (page 66).

Fraser, C, CA Donnelly, S Cauchemez, et al. (2009). 'Pandemic potential of a strain of influenza A (H1N1): early findings'. In: *Science* 324, pp. 1557–1561 (page 133).

Fu, YX (1994). 'A phylogenetic estimator of effective population size or mutation rate'. In: *Genetics* 136.2, pp. 685–692 (page 29).

Gavryushkina, A, D Welch and AJ Drummond (2013). 'Recursive algorithms for phylogenetic tree counting'. In: *Algorithms in Molecular Biology* 8, p. 26 (pages 25, 42).

Gavryushkina, A, D Welch, T Stadler and AJ Drummond (2014). 'Bayesian inference of sampled ancestor trees for epidemiology and fossil calibration'. In: *arXiv preprint arXiv:*1406.4573 (pages 42, 67, 111, 112).

Gelman, A and DB Rubin (1992). 'A single series from the Gibbs sampler provides a false sense of security'. In: *Bayesian statistics.* Ed. by J Bernardo, JO Berger, JO Dawid and AFM Smith. Vol. 4. Oxford: Oxford University Press, pp. 625–631 (page 140).

Gelman, A, GO Roberts and WR Gilks (1996). 'Efficient Metropolis jumping rules'. In: *Bayesian statistics.* Ed. by JM Bernardo, JO Berger, AP Dawid and AFM Smith. Vol. 5. Oxford: Oxford University Press, pp. 599–608 (page 89).

Gelman, A, J Carlin, H Stern and D Rubin (2004). *Bayesian data analysis.* 2nd edn. New York: Chapman & Hall/CRC (pages 10, 140).

Geman, S and D Geman (1984). 'Stochastic relaxation, Gibbs distribution, and the Bayesian restoration of images'. In: *IEEE Transactions on Pattern Analysis and Machine Intelligence* 6, pp. 721–741 (pages 16, 139).

Gernhard, T (2008). 'The conditioned reconstructed process'. In: *Journal of Theoretical Biology* 253.4, pp. 769–778 (page 38).

Geweke, J (1992). 'Evaluating the accuracy of sampling-based approaches to the calculation of posterior moments'. In: *Bayesian statistics*. Ed. by JM Bernardo, JO Berger, JO Dawid and AFM Smith. Vol. 4. Oxford: Oxford University Press, pp. 169–193 (page 140).

Gill, MS, P Lemey, NR Faria, et al. (2013). 'Improving Bayesian population dynamics inference: a coalescent-based model for multiple loci'. In: *Molecular Biology and Evolution* 30.3, pp. 713–724 (page 33).

Gillespie, JH (2001). 'Is the population size of a species relevant to its evolution?' In: *Evolution* 55.11, pp. 2161–2169 (page 29).

Goldman, N and Z Yang (1994). 'A codon-based model of nucleotide substitution for protein-coding DNA sequences'. In: *Molecular Biology and Evolution* 11, pp. 725–736 (page 51).

Goldstein, M (2006). 'Subjective Bayesian analysis: principles and practice'. In: *Bayesian Analysis* 1.3, pp. 403–420 (page 20).

Graham, M and J Kennedy (2010). 'A survey of multiple tree visualisation'. In: *Information Visualization* 9.4, pp. 235–252 (page 156).

Gray, RD and QD Atkinson (2003). 'Language-tree divergence times support the Anatolian theory of Indo-European origin'. In: *Nature* 426, pp. 435–439 (page 57).

Gray, RD, AJ Drummond and SJ Greenhill (2009). 'Language phylogenies reveal expansion pulses and pauses in Pacific settlement'. In: *Science* 323, pp. 479–483 (page 57).

Gray, RR, AJ Tatem, JA Johnson, et al. (2011). 'Testing spatiotemporal hypothesis of bacterial evolution using methicillin-resistant *Staphylococcus aureus* ST239 genome-wide data within a Bayesian framework'. In: *Molecular Biology and Evolution* 28.5, pp. 1593–1603 (page 75).

Green, P (1995). 'Reversible jump Markov chain Monte Carlo computation and Bayesian model determination'. In: *Biometrika* 82, pp. 711–732 (pages 16, 17).

Green, P, N Hjort and S Richardson (2003). *Highly structured stochastic systems*. Oxford: Oxford University Press (page 16).

Greenhill, SJ, AJ Drummond and RD Gray (2010). 'How accurate and robust are the phylogenetic estimates of Austronesian language relationships?' In: *PloS One* 5.3, e9573 (page 97).

Grenfell, BT, OG Pybus, JR Gog, et al. (2004). 'Unifying the epidemiological and evolutionary dynamics of pathogens'. In: *Science* 303, pp. 327–332 (page 74).

Griffiths, RC (1989). 'Genealogical-tree probabilities in the infinitely-many-site model'. In: *Journal of Mathematical Biology* 27.6, pp. 667–680 (page 31).

Griffiths, RC and P Marjoram (1996). 'Ancestral inference from samples of DNA sequences with recombination'. In: *Journal of Computational Biology* 3.4, pp. 479–502 (pages 10, 31).

Griffiths, RC and S Tavaré (1994). 'Sampling theory for neutral alleles in a varying environment'. In: *Philosophical Transactions of the Royal Society B: Biological Sciences* 344, pp. 403–410 (pages 29, 31).

Grummer, JA, RW Bryson and TW Reeder (2014). 'Species delimitation using Bayes factors: simulations and application to the *Sceloporus scalaris* species group (Squamata: Phrynosomatidae)'. In: *Systematic Biology* 63.2, pp. 119–133 (page 122).

Gu, X, YX Fu and WH Li (1995). 'Maximum likelihood estimation of the heterogeneity of substitution rate among nucleotide sites'. In: *Molecular Biology and Evolution* 12.4, pp. 546–557 (page 54).

Hailer, F, VE Kutschera, BM Hallström, et al. (2012). 'Nuclear genomic sequences reveal that polar bears are an old and distinct bear lineage'. In: *Science* 336, pp. 344–347. eprint: www.sciencemag.org/content/336/6079/344.full.pdf (page 121).

Harding, E (1971). 'The probabilities of rooted tree-shapes generated by random bifurcation'. In: *Advances in Applied Probability* 3, pp. 44–77 (page 36).

Harvey, PH and MD Pagel (1991). *The comparative method in evolutionary biology*. Oxford: Oxford University Press (page 73).

Hasegawa, M, H Kishino and T Yano (1985). 'Dating the human–ape splitting by a molecular clock of mitochondrial DNA'. In: *Journal of Molecular Evolution* 22, pp. 160–174 (pages 50, 104).

Hastings, W (1970). 'Monte Carlo sampling methods using Markov chains and their applications'. In: *Biometrika* 57, pp. 97–109 (pages 14, 16, 20, 31).

He, M, F Miyajima, P Roberts, et al. (2013). 'Emergence and global spread of epidemic healthcare-associated *Clostridium difficile*'. In: *Nature Genetics* 45.1, pp. 109–113 (page 134).

Heath, TA, JP Huelsenbeck and T Stadler (2014). 'The fossilized birth–death process for coherent calibration of divergence-time estimates'. In: *Proceedings of the National Academy of Sciences USA* 111.29, pp. 2957–2966 (pages 67, 112).

Heidelberger, P and PD Welch (1983). 'Simulation run length control in the presence of an initial transient'. In: *Operations Research* 31.6, pp. 1109–1144 (page 140).

Hein, J, M Schierup and C Wiuf (2004). *Gene genealogies, variation and evolution: a primer in coalescent theory*. Oxford: Oxford University Press (page 27).

Heled, J and RR Bouckaert (2013). 'Looking for trees in the forest: summary tree from posterior samples'. In: *BMC Evolutionary Biology* 13.1, art. 221 (pages 159, 160, 163).

Heled, J and AJ Drummond (2008). 'Bayesian inference of population size history from multiple loci'. In: *BMC Evolutionary Biology* 8.1, art. 289 (pages 6, 17, 32, 110, 130, 131, 151).

Heled, J and AJ Drummond (2010). 'Bayesian inference of species trees from multilocus data'. In: *Molecular Biology and Evolution* 27.3, pp. 570–580 (pages 7, 8, 100, 113, 116, 119, 122).

Heled, J and AJ Drummond (2012). 'Calibrated tree priors for relaxed phylogenetics and divergence time estimation'. In: *Systematic Biology* 61.1, pp. 138–149 (pages 66, 84, 110, 111, 127).

Heled, J and AJ Drummond (2013). 'Calibrated birth–death phylogenetic time-tree priors for Bayesian inference'. In: *arXiv preprint arXiv:*1311.4921 (pages 66, 110, 111, 127).

Hey, J (2010). 'Isolation with migration models for more than two populations'. In: *Molecular Biology and Evolution* 27.4, pp. 905–920 (page 8).

Hillis, D, T Heath and K St John (2005). 'Analysis and visualization of tree space'. In: *Systematic Biology* 54.3, pp. 471–482 (page 159).

Ho, SY and B Shapiro (2011). 'Skyline-plot methods for estimating demographic history from nucleotide sequences'. In: *Molecular Ecology Resources* 11.3, pp. 423–434 (page 97).

Ho, SYW, MJ Phillips, AJ Drummond and A Cooper (2005). 'Accuracy of rate estimation using relaxed-clock models with a critical focus on the early Metazoan radiation'. In: *Molecular Biology and Evolution* 22, pp. 1355–1363 (page 144).

Hoffman, J, S Grant, J Forcada and C Phillips (2011). 'Bayesian inference of a historical bottleneck in a heavily exploited marine mammal'. In: *Molecular Ecology* 20.19, pp. 3989–4008 (page 133).

Hoffman, MD and A Gelman (2014). 'The no-U-turn sampler: adaptively setting path lengths in Hamiltonian Monte Carlo'. In: *Journal of Machine Learning Research* 15, pp. 1593–1623 (page 156).

Höhna, S and AJ Drummond (2012). 'Guided tree topology proposals for Bayesian phylogenetic inference'. In: *Systematic Biology* 61.1, pp. 1–11 (pages 113, 150, 156, 159, 161).

Höhna, S, T Stadler, F Ronquist and T Britton (2011). 'Inferring speciation and extinction rates under different sampling schemes'. In: *Molecular Biology and Evolution* 28.9, pp. 2577–2589 (page 98).

Holder, MT, PO Lewis, DL Swofford and B Larget (2005). 'Hastings ratio of the LOCAL proposal used in Bayesian phylogenetics'. In: *Systematic Biology* 54, pp. 961–965 (page 16).

Holder, MT, J Sukumaran and PO Lewis (2008). 'A justification for reporting the majority-rule consensus tree in Bayesian phylogenetics'. In: *Systematic Biology* 57.5, pp. 814–821 (page 160).

Holmes, EC and BT Grenfell (2009). 'Discovering the phylodynamics of RNA viruses'. In: *PLoS Computational Biology* 5.10, e1000505 (page 74).

Holmes, EC, LQ Zhang, P Simmonds, AS Rogers and AJ Leigh Brown (1993). 'Molecular investigation of human immunodeficiency virus (HIV) infection in a patient of an HIV-infected surgeon'. In: *Journal of Infectious Diseases* 167.6, pp. 1411–1414 (page 31).

Hudson, RR (1987). 'Estimating the recombination parameter of a finite population model without selection'. In: *Genetics Research* 50.3, pp. 245–250 (page 29).

Hudson, RR (1990). 'Gene genealogies and the coalescent process'. In: *Oxford surveys in evolutionary biology*. Ed. by D Futuyma and J Antonovics. Vol. 7. Oxford: Oxford University Press, pp. 1–44 (pages 29, 71, 72).

Hudson, RR and NL Kaplan (1985). 'Statistical properties of the number of recombination events in the history of a sample of DNA sequences'. In: *Genetics* 111.1, pp. 147–164 (page 29).

Huelsenbeck, JP and F Ronquist (2001). 'MrBayes: Bayesian inference of phylogenetic trees'. In: *Bioinformatics* 17, pp. 754–755 (page 14).

Huelsenbeck, JP, B Larget and DL Swofford (2000). 'A compound Poisson process for relaxing the molecular clock'. In: *Genetics* 154, pp. 1879–1892 (page 62).

Huelsenbeck, JP, F Ronquist, R Nielsen and JP Bollback (2001). 'Bayesian inference of phylogeny and its impact on evolutionary biology'. In: *Science* 294, pp. 2310–2314 (pages 6, 111).

Huelsenbeck, JP, B Larget and ME Alfaro (2004). 'Bayesian phylogenetic model selection using reversible jump Markov chain Monte Carlo'. In: *Molecular Biology and Evolution* 21.6, pp. 1123–1133 (page 57).

Hurvich, CM and CL Tsai (1989). 'Regression and time series model selection in small samples'. In: *Biometrika* 76.2, pp. 297–307 (page 32).

Huson, DH and D Bryant (2006). 'Application of phylogenetic networks in evolutionary studies'. In: *Molecular Biology and Evolution* 23.2, pp. 254–267 (page 97).

Jackman, T, A Larson, KD Queiroz and J Losos (1999). 'Phylogenetic relationships and tempo of early diversification in *Anolis* lizards'. In: *Systematic Biology* 48.2, pp. 254–285 (page 157).

Jaynes, ET (2003). *Probability theory: the logic of science*. Cambridge: Cambridge University Press (pages 10, 19).

Jeffreys, H (1946). 'An invariant form for the prior probability in estimation problems'. In: *Proceedings of the Royal Society A: Mathematical and Physical Sciences* 186.1007, pp. 453–461 (page 14).

Jeffreys, H (1961). *Theory of probability*. 1st edn. London: Oxford University Press (page 14).

Jenkins, GM, A Rambaut, OG Pybus and EC Holmes (2002). 'Rates of molecular evolution in RNA viruses: a quantitative phylogenetic analysis'. In: *Journal of Molecular Evolution* 54.2, pp. 156–165 (pages 66, 152).

Jones, G (2011). 'Calculations for multi-type age-dependent binary branching processes'. In: *Journal of Mathematical Biology* 63.1, pp. 33–56 (page 39).

Jukes, T and C Cantor (1969). 'Evolution of protein molecules'. In: *Mammaliam protein metabolism*. Ed. by H Munro. New York: Academic Press, pp. 21–132 (pages 44, 46, 103).

Kass, R and A Raftery (1995). 'Bayes factors'. In: *Journal of the American Statistical Association* 90, pp. 773–795 (page 18).

Kass, RE, BP Carlin, A Gelman and RM Neal (1998). 'Markov chain Monte Carlo in practice: a roundtable discussion'. In: *The American Statistician* 52.2, pp. 93–100 (pages 140, 141).

Katoh, K and DM Standley (2013). 'MAFFT multiple sequence alignment software version 7: improvements in performance and usability'. In: *Molecular Biology and Evolution* 30.4, pp. 772–780 (page 9).

Katoh, K and DM Standley (2014). 'MAFFT: iterative refinement and additional methods'. In: *Methods in Molecular Biology* 1079, pp. 131–146 (page 9).

Katoh, K, K Misawa, K Kuma and T Miyata (2002). 'MAFFT: a novel method for rapid multiple sequence alignment based on fast Fourier transform'. In: *Nucleic Acids Research* 30.14, pp. 3059–3066 (page 9).

Keeling, MJ and P Rohani (2008). *Modeling infectious diseases in humans and animals.* Princeton, NJ: Princeton, University Press (page 34).

Kendall, DG (1948). 'On the generalized "birth-and-death" process'. In: *Annals of Mathematical Statistics* 19.1, pp. 1–15 (page 36).

Kendall, DG (1949). 'Stochastic processes and population growth'. In: *Journal of the Royal Statistical Society, Series B* 11.2, pp. 230–282 (page 73).

Kimura, M (1980). 'A simple model for estimating evolutionary rates of base substitutions through comparative studies of nucleotide sequences'. In: *Journal of Molecular Evolution* 16, pp. 111–120 (page 48).

Kingman, J (1982). 'The coalescent'. In: *Stochastic Processes and Their Applications* 13.3, pp. 235–248 (pages 6, 28).

Kishino, H, JL Thorne and WJ Bruno (2001). 'Performance of a divergence time estimation method under a probabilistic model of rate evolution'. In: *Molecular Biology and Evolution* 18.3, pp. 352–361 (page 62).

Knuth, D (1997). *The art of computer programming.* Vol. 2. *Seminumerical algorithms.* Reading, MA: Addison-Wesley (page 114).

Kouyos, RD, CL Althaus and S Bonhoeffer (2006). 'Stochastic or deterministic: what is the effective population size of HIV-1?' In: *Trends in Microbiology* 14.12, pp. 507–511 (page 29).

Kubatko, LS and JH Degnan (2007). 'Inconsistency of phylogenetic estimates from concatenated data under coalescence'. In: *Systematic Biology* 56.1, pp. 17–24 (page 116).

Kuhner, MK (2006). 'LAMARC 2.0: maximum likelihood and Bayesian estimation of population parameters'. In: *Bioinformatics* 22.6, pp. 768–770 (pages 10, 14).

Kuhner, MK, J Yamato and J Felsenstein (1995). 'Estimating effective population size and mutation rate from sequence data using Metropolis–Hastings sampling'. In: *Genetics* 140, pp. 1421–1430 (pages 6, 31).

Kuhner, MK, J Yamato and J Felsenstein (1998). 'Maximum likelihood estimation of population growth rates based on the coalescent'. In: *Genetics* 149, pp. 429–434 (pages 6, 31, 109).

Kuhner, MK, J Yamato and J Felsenstein (2000). 'Maximum likelihood estimation of recombination rates from population data'. In: *Genetics* 156.3, pp. 1393–1401 (pages 6, 10, 31).

Kühnert, D, CH Wu and AJ Drummond (2011). 'Phylogenetic and epidemic modeling of rapidly evolving infectious diseases'. In: *Infection, Genetics and Evolution* 11.8, pp. 1825–1841 (pages 8, 66, 75).

Kühnert, D, T Stadler, TG Vaughan and AJ Drummond (2014). 'Simultaneous reconstruction of evolutionary history and epidemiological dynamics from viral sequences with the birth–death SIR model'. In: *Journal of the Royal Society Interface* 11.94 (pages 40, 75, 130).

Kuo, L and B Mallick (1998). 'Variable selection for regression models'. In: *Sankhya B* 60, pp. 65–81 (page 17).

Lakner, C, P van der Mark, JP Huelsenbeck, B Larget and F Ronquist (2008). 'Efficiency of Markov chain Monte Carlo tree proposals in Bayesian phylogenetics'. In: *Systematic Biology* 57.1, pp. 86–103 (page 156).

Lambert, DM, PA Ritchie, CD Millar, et al. (2002). 'Rates of evolution in ancient DNA from Adelie penguins'. In: *Science* 295, pp. 2270–2273 (pages 31, 66, 129).

Laplace, P (1812). *Théorie analytique des probabilités*. Paris: Courcier (page 13).

Larget, B (2013). 'The estimation of tree posterior probabilities using conditional clade probability distributions'. In: *Systematic Biology* 62.4, pp. 501–511 (pages 159, 161).

Larget, B and D Simon (1999). 'Markov chain Monte Carlo algorithms for the Bayesian analysis of phylogenetic trees'. In: *Molecular Biology and Evolution* 16, pp. 750–759 (page 14).

Larkin, M, G Blackshields, NP Brown, et al. (2007). 'Clustal W and Clustal X version 2.0'. In: *Bioinformatics* 23, pp. 2947–2948 (page 9).

Lartillot, N and H Philippe (2006). 'Computing Bayes factors using thermodynamic integration'. In: *Systematic Biology* 55, pp. 195–207 (page 136).

Lartillot, N, T Lepage and S Blanquart (2009). 'PhyloBayes 3: a Bayesian software package for phylogenetic reconstruction and molecular dating'. In: *Bioinformatics* 25.17, pp. 2286–2288 (page 14).

Leaché, AD and MK Fujita (2010). 'Bayesian species delimitation in West African forest geckos (*Hemidactylus fasciatus*)'. In: *Proceedings of the Royal Society B: Biological Sciences* 277, pp. 3071–3077 (page 121).

Leaché, AD and B Rannala (2011). 'The accuracy of species tree estimation under simulation: a comparison of methods'. In: *Systematic Biology* 60.2, pp. 126–137 (page 119).

Leaché, AD, MK Fujita, VN Minin and RR Bouckaert (2014). 'Species delimitation using genome-wide SNP data'. In: *Systematic Biology*, syu018 (page 126).

Leitner, T and W Fitch (1999). 'The phylogenetics of known transmission histories'. In: *The evolution of HIV*. Ed. by KA Crandall. Baltimore, MD: Johns Hopkins University Press, pp. 315–345 (page 41).

Lemey, P, OG Pybus, B Wang, et al. (2003). 'Tracing the origin and history of the HIV-2 epidemic'. In: *Proceedings of the National Academy of Sciences USA* 100.11, pp. 6588–6592 (page 32).

Lemey, P, OG Pybus, A Rambaut, et al. (2004). 'The molecular population genetics of HIV-1 group O'. In: *Genetics* 167.3, pp. 1059–1068 (pages 32, 152).

Lemey, P, A Rambaut, AJ Drummond and MA Suchard (2009a). 'Bayesian phylogeography finds its roots'. In: *PLoS Computational Biology* 5.9, e1000520 (pages 17, 69, 71, 72, 76, 134).

Lemey, P, MA Suchard and A Rambaut (2009b). 'Reconstructing the initial global spread of a human influenza pandemic: a Bayesian spatial–temporal model for the global spread of H1N1pdm'. In: *PLoS Currents* RRN1031 (pages 71, 129, 134).

Lemey, P, A Rambaut, JJ Welch and MA Suchard (2010). 'Phylogeography takes a relaxed random walk in continuous space and time'. In: *Molecular Biology and Evolution* 27.8, pp. 1877–1885 (pages 73, 134).

Lemey, P, A Rambaut, T Bedford, et al. (2014). 'Unifying viral genetics and human transportation data to predict the global transmission dynamics of human influenza H3N2'. In: *PLoS Pathogens* 10.2, e1003932 (page 135).

Leonard, J, R Wayne, J Wheeler, et al. (2002). 'Ancient DNA evidence for Old World origin of New World dogs'. In: *Science* 298, p. 1613 (page 31).

Lepage, T, D Bryant, H Philippe and N Lartillot (2007). 'A general comparison of relaxed molecular clock models'. In: *Molecular Biology and Evolution* 24.12, pp. 2669–2680 (page 63).

Leventhal, GE, H Guenthard, S Bonhoeffer and T Stadler (2014). 'Using an epidemiological model for phylogenetic inference reveals density-dependence in HIV transmission'. In: *Molecular Biology and Evolution* 31.1, pp. 6–17 (pages 34, 40, 75).

Levinson, G and GA Gutman (1987). 'High frequencies of short frameshifts in poly-CA/TG tandem repeats borne by bacteriophage M13 in *Escherichia coli* K-12'. In: *Nucleic Acids Research* 15.13, pp. 5323–5338 (page 52).

Lewis, PO (2001). 'A likelihood approach to estimating phylogeny from discrete morphological character data'. In: *Systematic Biology* 50.6, pp. 913–925 (page 57).

Lewis, PO, MT Holder and KE Holsinger (2005). 'Polytomies and Bayesian phylogenetic inference.' In: *Systematic Biology* 54.2, pp. 241–253 (page 14).

Li, S, D Pearl and H Doss (2000). 'Phylogenetic tree construction using Markov chain Monte Carlo'. In: *Journal of the American Statistical Association* 95, pp. 493–508 (page 14).

Li, WLS and AJ Drummond (2012). 'Model averaging and Bayes factor calculation of relaxed molecular clocks in Bayesian phylogenetics'. In: *Molecular Biology and Evolution* 29.2, pp. 751–761 (pages 63, 145).

Liu, L (2008). 'BEST: Bayesian estimation of species trees under the coalescent model'. In: *Bioinformatics* 24.21, pp. 2542–2543 (page 119).

Liu, L, DK Pearl, RT Brumfield and SV Edwards (2008). 'Estimating species trees using multiple-allele DNA sequence data'. In: *Evolution* 62.8, pp. 2080–2091 (page 7).

Liu, L, L Yu, L Kubatko, DK Pearl and SV Edwards (2009a). 'Coalescent methods for estimating phylogenetic trees'. In: *Molecular Phylogenetics and Evolution* 53.1, pp. 320–328 (page 7).

Liu, L, L Yu, DK Pearl and SV Edwards (2009b). 'Estimating species phylogenies using coalescence times among sequences'. In: *Systematic Biology* 58.5, pp. 468–477 (page 7).

Loreille, O, L Orlando, M Patou-Mathis, et al. (2001). 'Ancient DNA analysis reveals divergence of the cave bear, *Ursus spelaeus*, and brown bear, *Ursus arctos*, lineages'. In: *Current Biology* 11.3, pp. 200–203 (page 31).

Lunn, DJ, A Thomas, N Best and D Spiegelhalter (2000). 'WinBUGS – a Bayesian modelling framework: concepts, structure, and extensibility'. In: *Statistics and Computing* 10.4, pp. 325–337 (page 156).

Lunn, D, D Spiegelhalter, A Thomas and N Best (2009). 'The BUGS project: evolution, critique and future directions'. In: *Statistics in Medicine* 28.25, pp. 3049–3067 (page 156).

Lunter, G, I Miklos, AJ Drummond, JL Jensen and J Hein (2005). 'Bayesian coestimation of phylogeny and sequence alignment'. In: *BMC Bioinformatics* 6, art. 83 (pages 4, 9, 99).

MacKay, DJ (2003). *Information theory, inference and learning algorithms*. Cambridge: Cambridge University Press (page 10).

Maddison, DR and WP Maddison (2005). *MacClade 4.08*. Sunderland, MA: Sinauer Associates (page 68).

Maddison, WP (2007). 'Estimating a binary character's effect on speciation and extinction'. In: *Systematic Biology* 56.5, pp. 701–710 (pages 39, 69).

Matschiner, M and RR Bouckaert (2013). 'A rough guide to CladeAge'. In: *BEAST 2 wiki* (pages 66, 111).

Matsumoto, M and T Nishimura (1998). 'Mersenne twister: a 623-dimensionally equidistributed uniform pseudo-random number generator'. In: *ACM Transactions on Modeling and Computer Simulation* 8.1, pp. 3–30 (page 114).

Mau, B and M Newton (1997). 'Phylogenetic inference for binary data on dendrograms using Markov chain Monte Carlo'. In: *Journal of Computational and Graphical Statistics* 6, pp. 122–131 (page 14).

Mau, B, MA Newton and B Larget (1999). 'Bayesian phylogenetic inference via Markov chain Monte Carlo methods'. In: *Biometrics* 55.1, pp. 1–12 (pages 14, 157).

McCormack, JE, J Heled, KS Delaney, AT Peterson and LL Knowles (2011). 'Calibrating divergence times on species trees versus gene trees: implications for speciation history of *Aphelocoma* jays'. In: *Evolution* 65.1, pp. 184–202 (page 165).

Meredith, R, J Janečka, J Gatesy, et al. (2011). 'Impacts of the Cretaceous terrestrial revolution and KPg extinction on mammal diversification'. In: *Science* 334, pp. 521–524 (page 39).

Metropolis, N, A Rosenbluth, M Rosenbluth, A Teller and E Teller (1953). 'Equations of state calculations by fast computing machines'. In: *Journal of Chemistry and Physics* 21, pp. 1087–1092 (pages 14, 20, 31).

Minin, VN, EW Bloomquist and MA Suchard (2008). 'Smooth skyride through a rough skyline: Bayesian coalescent-based inference of population dynamics'. In: *Molecular Biology and Evolution* 25.7, pp. 1459–1471 (pages 33, 108).

Molina, J, M Sikora, N Garud, et al. (2011). 'Molecular evidence for a single evolutionary origin of domesticated rice'. In: *Proceedings of the National Academy of Sciences USA* 108.20, pp. 8351–8356 (page 121).

Mollentze, N, LH Nel, S Townsend, et al. (2014). 'A Bayesian approach for inferring the dynamics of partially observed endemic infectious diseases from space-time-genetic data'. In: *Proceedings of the Royal Society B: Biological Sciences* 281.1782, p. 20133251 (page 75).

Monjane, AL, GW Harkins, DP Martin, et al. (2011). 'Reconstructing the history of Maize streak virus strain A dispersal to reveal diversification hot spots and its origin in southern Africa'. In: *Journal of Virology* 85.18, pp. 9623–9636 (page 135).

Mooers, A and S Heard (1997). 'Inferring evolutionary process from phylogenetic tree shape'. In: *Quarterly Review of Biology* 72, pp. 31–54 (page 107).

Moran, PAP (1962). *The statistical processes of evolutionary theory*. Oxford: Clarendon Press (page 28).

Moran, PAP (1958). 'Random processes in genetics'. In: *Mathematical Proceedings of the Cambridge Philosophical Society* 54, pp. 60–71 (page 28).

Morlon, H, T Parsons and J Plotkin (2011). 'Reconciling molecular phylogenies with the fossil record'. In: *Proceedings of the National Academy of Sciences USA* 108.39, pp. 16 327–16 332 (page 39).

Mourier, T, SY Ho, MTP Gilbert, E Willerslev and L Orlando (2012). 'Statistical guidelines for detecting past population shifts using ancient DNA'. In: *Molecular Biology and Evolution* 29.9, pp. 2241–2251 (page 97).

Muse, S and B Gaut (1994). 'A likelihood approach for comparing synonymous and nonsynonymous nucleotide substitution rates, with applications to the chloroplast genome'. In: *Molecular Biology and Evolution* 11, pp. 715–725 (page 51).

Mutreja, A, DW Kim, NR Thomson, et al. (2011). 'Evidence for several waves of global transmission in the seventh cholera pandemic'. In: *Nature* 477.7365, pp. 462–465 (page 129).

Nagalingum, N, C Marshall, T Quental, et al. (2011). 'Recent synchronous radiation of a living fossil'. In: *Science* 334, pp. 796–799 (page 96).

Nee, SC (2001). 'Inferring speciation rates from phylogenies'. In: *Evolution* 55.4, pp. 661–668 (page 107).

Nee, SC, EC Holmes, RM May and PH Harvey (1994a). 'Extinction rates can be estimated from molecular phylogenies'. In: *Philosophical Transactions of the Royal Society B: Biological Sciences* 344.1307, pp. 77–82 (page 107).

Nee, SC, RM May and PH Harvey (1994b). 'The reconstructed evolutionary process'. In: *Philosophical Transactions of the Royal Society B: Biological Sciences* 344, pp. 305–311 (page 39).

Nee, SC, EC Holmes, A Rambaut and PH Harvey (1995). 'Inferring population history from molecular phylogenies'. In: *Philosophical Transactions of the Royal Society B: Biological Sciences* 349.1327, pp. 25–31 (page 29).

Newton, M and A Raftery (1994). 'Approximate Bayesian inference with the weighted likelihood bootstrap'. In: *Journal of the Royal Statistical Society, Series B* 56, pp. 3–48 (page 18).

Nicholls, GK and RD Gray (2006). 'Quantifying uncertainty in a stochastic model of vocabulary evolution'. In: *Phylogenetic methods and the prehistory of languages*. Ed. by P Forster and C Renfrew. Cambridge: McDonald Institute for Archaeological Research, pp. 161–171 (page 57).

Nicholls, GK and RD Gray (2008). 'Dated ancestral trees from binary trait data and their application to the diversification of languages'. In: *Journal of the Royal Statistical Society, Series B* 70.3, pp. 545–566 (page 115).

Nielsen, R and Z Yang (1998). 'Likelihood models for detecting positively selected amino acid sites and applications to the HIV-1 envelope gene'. In: *Genetics* 148, pp. 929–936 (page 111).

Notredame, C, DG Higgins and J Heringa (2000). 'T-coffee: a novel method for fast and accurate multiple sequence alignment'. In: *Journal of Molecular Biology* 302.1, pp. 205 –217 (page 9).

Novák, A, I Miklós, R Lyngsø and J Hein (2008). 'StatAlign: an extendable software package for joint Bayesian estimation of alignments and evolutionary trees'. In: *Bioinformatics* 24.20, pp. 2403–2404 (pages 9, 99).

Nylander, JAA, JC Wilgenbusch, DL Warren and DL Swofford (2008). 'AWTY (are we there yet?): a system for graphical exploration of MCMC convergence in Bayesian phylogenetics'. In: *Bioinformatics* 24.4, pp. 581–583 (pages 141, 156).

Ohta, T and M Kimura (1973). 'A model of mutation appropriate to estimate the number of electrophoretically detectable alleles in a finite population'. In: *Genetics* 22.2, pp. 201–204 (page 52).

Opgen-Rhein, R, L Fahrmeir and K Strimmer (2005). 'Inference of demographic history from genealogical trees using reversible jump Markov chain Monte Carlo'. In: *BMC Evolutionary Biology* 5.1, art. 6 (page 32).

Owen, M and JS Provan (2011). 'A fast algorithm for computing geodesic distances in tree space'. In: *IEEE/ACM Transactions on Computational Biology and Bioinformatics* 8.1, pp. 2–13 (page 154).

Pagel, MD and A Meade (2004). 'A phylogenetic mixture model for detecting pattern-heterogeneity in gene sequence or character-state data'. In: *Systematic Biology* 53.4, pp. 571–581 (page 14).

Palacios, JA and VN Minin (2012). 'Integrated nested Laplace approximation for Bayesian nonparametric phylodynamics.' In: *Proceedings of the 28th Conference on Uncertainty in Artificial Intelligence*. Ed. by N de Freitas and K Murphy. Sterling, VA: AUAI Press, pp. 726–735 (page 33).

Palacios, JA and VN Minin (2013). 'Gaussian process-based Bayesian nonparametric inference of population trajectories from gene genealogies'. In: *Biometrics* 69.1, pp. 8–18 (page 33).

Palmer, D, J Frater, R Phillips, AR McLean and G McVean (2013). 'Integrating genealogical and dynamical modelling to infer escape and reversion rates in HIV epitopes'. In: *Proceedings of the Royal Society B: Biological Sciences* 280.1762, art. 2013.0696 (pages 69, 75).

Pamilo, P and M Nei (1988). 'Relationships between gene trees and species trees'. In: *Molecular Biology and Evolution* 5.5, pp. 568–83 (pages 41, 116).

Penny, D, BJ McComish, MA Charleston and MD Hendy (2001). 'Mathematical elegance with biochemical realism: the covarion model of molecular evolution'. In: *Journal of Molecular Evolution* 53.6, pp. 711–723 (page 57).

Pereira, L, F Freitas, V Fernandes, et al. (2009). 'The diversity present in 5140 human mitochondrial genomes'. In: *American Journal of Human Genetics* 84.5, pp. 628–640 (page 102).

Popinga, A, TG Vaughan, T Stadler and AJ Drummond (2015). 'Inferring epidemiological dynamics with Bayesion coalescent inference: the merits of deterministic and Stochastic models'. In: *Genetics* 199.2, pp. 595–607 (page 36).

Posada, D (2008). 'jModelTest: phylogenetic model averaging'. In: *Molecular Biology and Evolution* 25.7, pp. 1253–1256 (page 145).

Posada, D and KA Crandall (1998). 'Modeltest: testing the model of DNA substitution.' In: *Bioinformatics* 14.9, pp. 817–818 (page 145).

Procter, JB, J Thompson, I Letunic, et al. (2010). 'Visualization of multiple alignments, phylogenies and gene family evolution'. In: *Nature Methods* 7, S16–S25 (page 156).

Pybus, OG and A Rambaut (2002). 'GENIE: estimating demographic history from molecular phylogenies'. In: *Bioinformatics* 18.10, pp. 1404–1405 (page 29).

Pybus, OG and A Rambaut (2009). 'Evolutionary analysis of the dynamics of viral infectious disease'. In: *Nature Reviews Genetics* 10.8, pp. 540–550 (pages 8, 75).

Pybus, OG, A Rambaut and PH Harvey (2000). 'An integrated framework for the inference of viral population history from reconstructed genealogies'. In: *Genetics* 155, pp. 1429–1437 (pages 29, 32).

Pybus, OG, MA Charleston, S Gupta, et al. (2001). 'The epidemic behavior of the hepatitis C virus'. In: *Science* 292, pp. 2323–2325 (page 29).

Pybus, OG, AJ Drummond, T Nakano, BH Robertson and A Rambaut (2003). 'The epidemiology and iatrogenic transmission of hepatitis C virus in Egypt: a Bayesian coalescent approach'. In: *Molecular Biology and Evolution* 20.3, pp. 381–387 (pages 29, 32, 106).

Pybus, OG, E Barnes, R Taggest, et al. (2009). 'Genetic history of hepatitis C virus in East Asia'. In: *Journal of Virology* 83.2, pp. 1071–1082 (page 33).

Pybus, OG, MA Suchard, P Lemey, et al. (2012). 'Unifying the spatial epidemiology and molecular evolution of emerging epidemics'. In: *Proceedings of the National Academy of Sciences USA* 109.37, pp. 15 066–15 071 (page 135).

Pyron, RA (2011). 'Divergence time estimation using fossils as terminal taxa and the origins of Lissamphibia'. In: *Systematic Biology*, syr047 (page 57).

Rabosky, D (2007). 'Likelihood methods for detecting temporal shifts in diversification rates'. In: *Evolution* 60.6, pp. 1152–1164 (page 39).

Raftery, AE and SM Lewis (1992). '[Practical Markov chain Monte Carlo]: comment: one long run with diagnostics: implementation strategies for Markov chain Monte Carlo'. In: *Statistical Science* 7.4, pp. 493–497 (page 140).

Raftery, A, M Newton, J Satagopan and P Krivitsky (2007). 'Estimating the integrated likelihood via posterior simulation using the harmonic mean identity'. In: *Bayesian statistics*. Ed. by JM Bernardo, MJ Bayarri, JO Berger et al. Vol. 8. Oxford: Oxford University Press, pp. 1–45 (page 18).

Rambaut, A (2000). 'Estimating the rate of molecular evolution: incorporating non-contemporaneous sequences into maximum likelihood phylogenies'. In: *Bioinformatics* 16, pp. 395–399 (pages 31, 66).

Rambaut, A (2010). *Path-O-Gen: temporal signal investigation tool. Version 1.3* (page 149).

Rambaut, A and AJ Drummond (2014). *Tracer v1.6* http://tree.bio.ed.ac.uk/software/tracer/ (pages 130, 137).

Rannala, B and Z Yang (1996). 'Probability distribution of molecular evolutionary trees: a new method of phylogenetic inference'. In: *Journal of Molecular Evolution* 43, pp. 304–311 (page 160).

Rannala, B and Z Yang (2007). 'Inferring speciation times under an episodic molecular clock'. In: *Systematic Biology* 56.3, pp. 453–466 (pages 62, 63).

Rasmussen, DA, O Ratmann and K Koelle (2011). 'Inference for nonlinear epidemiological models using genealogies and time series'. In: *PLoS Computational Biology* 7.8, e1002136 (page 75).

Redelings, BD and MA Suchard (2005). 'Joint Bayesian estimation of alignment and phylogeny'. In: *Systematic Biology* 54.3, pp. 401–418 (pages 4, 9, 99, 137).

Redelings, BD and MA Suchard (2007). 'Incorporating indel information into phylogeny estimation for rapidly emerging pathogens'. In: *BMC Evolutionary Biology* 7, art. 40 (page 99).

Reid, NM, SM Hird, JM Brown, et al. (2013). 'Poor fit to the multispecies coalescent is widely detectable in empirical data'. In: *Systematic Biology*, syt057 (page 116).

Richard, GF and F Pâques (2000). 'Mini- and microsatellite expansions: the recombination connection'. In: *EMBO Report* 1.2, pp. 122–126 (page 52).

Robinson, D and LR Foulds (1981). 'Comparison of phylogenetic trees'. In: *Mathematical Biosciences* 53.1, pp. 131–147 (page 161).

Rodrigo, AG and J Felsenstein (1999). 'Coalescent approaches to HIV population genetics'. In: *The evolution of HIV*. Ed. by KA Crandall. Baltimore, MD: Johns Hopkins University Press, pp. 233–272 (pages 31, 33).

Rodrigo, AG, EG Shpaer, EL Delwart, et al. (1999). 'Coalescent estimates of HIV-1 generation time in vivo'. In: *Proceedings of the National Academy Sciences USA* 96.5, pp. 2187–2191 (page 31).

Rodrigue, N, H Philippe and N Lartillot (2008). 'Uniformization for sampling realizations of Markov processes: applications to Bayesian implementations of codon substitution models'. In: *Bioinformatics* 24.1, pp. 56–62 (page 72).

Romero-Severson, E, H Skar, I Bulla, J Albert and T Leitner (2014). 'Timing and order of transmission events is not directly reflected in a pathogen phylogeny'. In: *Molecular Biology and Evolution* 31.9, pp. 2472–2482 (page 41).

Ronquist, F and JP Huelsenbeck (2003). 'MrBayes 3: Bayesian phylogenetic inference under mixed models'. In: *Bioinformatics* 19.12, pp. 1572–1574 (page 23).

Ronquist, F, M Teslenko, P van der Mask, et al. (2012a). 'MrBayes 3.2: efficient Bayesian phylogenetic inference and model choice across a large model space'. In: *Systematic Biology* 61.3, pp. 539–542 (page 74).

Ronquist, F, S Klopfstein, L Vilhelmsen, et al. (2012b). 'A total-evidence approach to dating with fossils, applied to the early radiation of the Hymenoptera'. In: *Systematic Biology* 61.6, pp. 973–999 (pages 65, 66).

Rosenberg, MS (2009). *Sequence alignment: methods, models, concepts, and strategies*. Berkeley, CA: University of California Press (page 8).

Rosenberg, M, S Subramanian and S Kumar (2003). 'Patterns of transitional mutation biases within and among mammalian genomes'. In: *Molecular Biology and Evolution* 20.6, pp. 988–993 (pages 84, 102).

Roure, B, D Baurain and H Philippe (2013). 'Impact of missing data on phylogenies inferred from empirical phylogenomic data sets'. In: *Molecular Biology and Evolution* 30.1, pp. 197–214 (page 99).

Sanmartín, I, P van der Mark and F Ronquist (2008). 'Inferring dispersal: a Bayesian approach to phylogeny-based island biogeography, with special reference to the Canary Islands'. In: *Journal of Biogeography* 35.3, pp. 428–449 (page 68).

Sarich, VM and AC Wilson (1967). 'Immunological time scale for hominid evolution'. In: *Science* 158, pp. 1200–1203 (pages 58, 60).

Sato, A, C O'hUigin, F Figueroa, et al. (1999). 'Phylogeny of Darwin's finches as revealed by mtDNA sequences'. In: *Proceedings of the National Academy of Sciences USA* 96, pp. 5101–5106 (page 116).

Seo, TK, JL Thorne, M Hasegawa and H Kishino (2002). 'A viral sampling design for testing the molecular clock and for estimating evolutionary rates and divergence times'. In: *Bioinformatics* 18.1, pp. 115–123 (page 7).

Shankarappa, R, JB Margolick, SJ Gange, et al. (1999). 'Consistent viral evolutionary changes associated with the disease progression of human immunodeficiency virus type 1 infection'. In: *Journal of Virology* 73, pp. 10489–10502 (page 31).

Shapiro, B, AJ Drummond, A Rambaut, et al. (2004). 'Rise and fall of the Beringian steppe bison.' In: *Science* 306, pp. 1561–1565 (pages 66, 129).

Shapiro, B, A Rambaut and AJ Drummond (2006). 'Choosing appropriate substitution models for the phylogenetic analysis of protein-coding sequences'. In: *Molecular Biology and Evolution* 23.1, pp. 7–9 (pages 100, 104, 146).

Silva, E de, N Ferguson and C Fraser (2012). 'Inferring pandemic growth rates from sequence data'. In: *Journal of the Royal Society Interface* 9.73, pp. 1797–1808 (pages 97, 132).

Sjödin, P, I Kaj, S Krone, M Lascoux and M Nordborg (2005). 'On the meaning and existence of an effective population size'. In: *Genetics* 169.2, pp. 1061–1070 (page 29).

Slatkin, M (1991). 'Inbreeding coefficients and coalescence times'. In: *Genetic Research* 58.2, pp. 167–175 (page 29).

Slatkin, M and WP Maddison (1989). 'A cladistic measure of gene flow inferred from the phylogenies of alleles'. In: *Genetics* 123.3, pp. 603–613 (page 68).

Smith, BJ (2007). 'boa: an R package for MCMC output convergence assessment and posterior inference'. In: *Journal of Statistical Software* 21.11, pp. 1–37 (page 140).

Smith, GP (1976). 'Evolution of repeated DNA sequences by unequal crossover'. In: *Science* 191, pp. 528–535 (page 52).

Smith, SA and MJ Donoghue (2008). 'Rates of molecular evolution are linked to life history in flowering plants'. In: *Science* 322, pp. 86–89 (page 96).

Stadler, T (2009). 'On incomplete sampling under birth–death models and connections to the sampling-based coalescent'. In: *Journal of Theoretical Biology* 261.1, pp. 58–66 (pages 38, 108).

Stadler, T (2010). 'Sampling-through-time in birth–death trees'. In: *Journal of Theoretical Biology* 267.3, pp. 396–404 (pages 36, 40, 75).

Stadler, T (2011). 'Mammalian phylogeny reveals recent diversification rate shifts'. In: *Proceedings of the National Academy of Sciences USA* 108.15, pp. 6187–6192 (page 39).

Stadler, T (2013a). 'Recovering speciation and extinction dynamics based on phylogenies'. In: *Journal of Evolutionary Biology* 26.6, pp. 1203–1219 (page 39).

Stadler, T (2013b). 'How can we improve accuracy of macroevolutionary rate estimates?' In: *Systematic Biology* 62.2, pp. 321–329 (pages 37, 38).

Stadler, T and S Bonhoeffer (2013). 'Uncovering epidemiological dynamics in heterogeneous host populations using phylogenetic methods'. In: *Philosophical Transactions of the Royal Society B: Biological Sciences* 368.1614 (pages 40, 70, 73, 75).

Stadler, T, R Kouyos, V von Wyl, et al. (2012). 'Estimating the basic reproductive number from viral sequence data'. In: *Molecular Biology and Evolution* 29, pp. 347–357 (page 40).

Stadler, T, D Kühnert, S Bonhoeffer and AJ Drummond (2013). 'Birth–death skyline plot reveals temporal changes of epidemic spread in HIV and HCV'. In: *Proceedings of the National Academy of Sciences USA* 110.1, pp. 228–233 (pages 39, 40, 75, 108, 130, 132).

Stadler, T, TG Vaughan, A Gavruskin, et al. (2015). 'How well can the exponential-growth coalescent approximate constant rate birth–death population dynamics?' In: *Proceedings of the Royal Society B: Biological Sciences,* in press (pages 36, 38).

Stan Development Team (2014). *Stan: A C++ Library for probability and sampling. Version 2.4* (page 156).

Steel, M (2005). 'Should phylogenetic models be trying to "fit an elephant"?' In: *Trends in Genetics* 21.6, pp. 307–309 (page 149).

Steel, M and A Mooers (2010). 'The expected length of pendant and interior edges of a Yule tree'. In: *Applied Mathematics Letters* 23.11, pp. 1315–1319 (page 110).

Stephens, M and P Donnelly (2000). 'Inference in molecular population genetics'. In: *Journal of the Royal Statistical Society. Series B,* 62.4, pp. 605–655 (page 31).

Stewart, WJ (1994). *Introduction to the numerical solution of Markov chains.* Vol. 41. Princeton, NJ: Princeton University Press (pages 12, 45).

Stockham, C, LS Wang and T Warnow (2002). 'Statistically based postprocessing of phylogenetic analysis by clustering'. In: *Bioinformatics* 18.suppl 1, S285–S293 (page 161).

Strimmer, K and OG Pybus (2001). 'Exploring the demographic history of DNA sequences using the generalized skyline plot'. In: *Molecular Biology and Evolution* 18.12, pp. 2298–2305 (page 32).

Suchard, MA and A Rambaut (2009). 'Many-core algorithms for statistical phylogenetics'. In: *Bioinformatics* 25.11, pp. 1370–1376 (pages 71, 114).

Suchard, MA and BD Redelings (2006). 'BAli-Phy: simultaneous Bayesian inference of alignment and phylogeny'. In: *Bioinformatics* 22.16, pp. 2047–2048 (pages 9, 99).

Suchard, MA, RE Weiss and JS Sinsheimer (2001). 'Bayesian selection of continuous-time Markov chain evolutionary models'. In: *Molecular Biology and Evolution* 18, pp. 1001–1013 (page 147).

Suchard, MA, RE Weiss and JS Sinsheimer (2003). 'Testing a molecular clock without an outgroup: derivations of induced priors on branch length restrictions in a Bayesian framework'. In: *Systematic Biology* 52, pp. 48–54 (page 99).

Sullivan, J and DL Swofford (2001). 'Should we use model-based methods for phylogenetic inference when we known that assumptions about among-site rate variation and nucleotide substitution pattern are violated?' In: *Systematic Biology* 50, pp. 723–729 (page 54).

Sullivan, J, DL Swofford and GJ Naylor (1999). 'The effect of taxon sampling on estimating rate heterogeneity parameters of maximum-likelihood models'. In: *Molecular Biology and Evolution* 16, pp. 1347–1356 (page 54).

Swofford, DL (2003). *PAUP*: phylogenetic analysis using parsimony (* and other methods). Version 4*. Sunderland, MA: Sinauer Associates (page 68).

Tajima, F (1983). 'Evolutionary relationship of DNA sequences in finite populations'. In: *Genetics* 105.2, pp. 437–460 (pages 29, 43).

Tajima, F (1989). 'DNA polymorphism in a subdivided population: the expected number of segregating sites in the two-subpopulation model'. In: *Genetics* 123.1, pp. 229–240 (page 29).

Takahata, N (1989). 'Gene genealogy in three related populations: consistency probability between gene and population trees'. In: *Genetics* 122.4, pp. 957–966 (page 29).

Tamura, K and M Nei (1993). 'Estimation of the number of nucleotide substitutions in the control region of mitochondrial DNA in humans and chimpanzees'. In: *Molecular Biology and Evolution* 10, pp. 512–526 (page 103).

Teixeira, S, EA Serrão and S Arnaud-Haond (2012). 'Panmixia in a fragmented and unstable environment: the hydrothermal shrimp *Rimicaris exoculata* disperses extensively along the mid-Atlantic Ridge'. In: *PloS One* 7.6, e38521 (page 133).

Thorne, JL and H Kishino (2002). 'Divergence time and evolutionary rate estimation with multilocus data.' In: *Systematic Biology* 51.5, pp. 689–702 (page 62).

Thorne, JL, H Kishino and IS Painter (1998). 'Estimating the rate of evolution of the rate of molecular evolution'. In: *Molecular Biology and Evolution* 15.12, pp. 1647–1657 (page 62).

Vaughan, TG and AJ Drummond (2013). 'A stochastic simulator of birth–death master equations with application to phylodynamics'. In: *Molecular Biology and Evolution* 30.6, pp. 1480–1493 (pages 43, 138).

Vaughan, TG, D Kühnert, A Popinga, D Welch and AJ Drummond (2014). 'Efficient Bayesian inference under the structured coalescent'. In: *Bioinformatics*, btu201 (pages 14, 71, 72).

Vijaykrishna, D, GJ Smith, OG Pybus, et al. (2011). 'Long-term evolution and transmission dynamics of swine influenza A virus'. In: *Nature* 473.7348, pp. 519–522 (page 129).

Volz, EM (2012). 'Complex population dynamics and the coalescent under neutrality'. In: *Genetics* 190.1, pp. 187–201 (pages 35, 69, 75).

Volz, EM, SL Kosakovsky Pond, MJ Ward, AJ Leigh Brown and SDW Frost (2009). 'Phylodynamics of infectious disease epidemics'. In: *Genetics* 183.4, pp. 1421–1430 (page 75).

Volz, EM, K Koelle and T Bedford (2013). 'Viral phylodynamics'. In: *PLoS Computational Biology* 9.3, e1002947 (pages 8, 75).

Waddell, P and D Penny (1996). 'Evolutionary trees of apes and humans from DNA sequences'. In: *Handbook of symbolic evolution*. Ed. by AJ Lock and CR Peters. Oxford: Clarendon Press, pp. 53–73 (page 54).

Wakeley, J and O Sargsyan (2009). 'Extensions of the coalescent effective population size'. In: *Genetics* 181.1, pp. 341–345 (page 29).

Wallace, R, H HoDac, R Lathrop and W Fitch (2007). 'A statistical phylogeography of influenza A H5N1'. In: *Proceedings of the National Academy of Sciences USA* 104.11, pp. 4473–4478 (page 68).

Welch, D (2011). 'Is network clustering detectable in transmission trees?' In: *Viruses* 3.6, pp. 659–676 (page 75).

Whidden, C, I Matsen and A Frederick (2014). 'Quantifying MCMC exploration of phylogenetic tree space'. In: *arXiv preprint arXiv:*1405.2120 (pages 150, 156).

Wiens, JJ and DS Moen (2008). 'Missing data and the accuracy of Bayesian phylogenetics'. In: *Journal of Systematics and Evolution* 46.3, pp. 307–314 (page 99).

Wilson, AC and VM Sarich (1969). 'A molecular time scale for human evolution'. In: *Proceedings of the National Academy of Sciences USA* 63.4, pp. 1088–1093 (page 60).

Wilson, IJ and DJ Balding (1998). 'Genealogical inference from microsatellite data'. In: *Genetics* 150.1, pp. 499–510 (page 14).

Wolinsky, S, B Korber, A Neumann, et al. (1996). 'Adaptive evolution of human immuno-deficiency virus type-1 during the natural course of infection'. In: *Science* 272, pp. 537–542 (page 31).

Wong, KM, MA Suchard and JP Huelsenbeck (2008). 'Alignment uncertainty and genomic analysis'. In: *Science* 319, pp. 473–476 (page 9).

Worobey, M, M Gemmel, DE Teuwen, et al. (2008). 'Direct evidence of extensive diversity of HIV-1 in Kinshasa by 1960'. In: *Nature* 455.7213, pp. 661–664 (page 32).

Worobey, M, P Telfer, S Souquière, et al. (2010). 'Island biogeography reveals the deep history of SIV'. In: *Science* 329, p. 1487 (page 129).

Wright, S (1931). 'Evolution in Mendelian populations'. In: *Genetics* 16.2, pp. 97–159 (page 28).

Wu, CH and AJ Drummond (2011). 'Joint inference of microsatellite mutation models, population history and genealogies using transdimensional Markov chain Monte Carlo'. In: *Genetics* 188.1, pp. 151–164 (pages 17, 52).

Wu, CH, MA Suchard and AJ Drummond (2013). 'Bayesian selection of nucleotide substitution models and their site assignments'. In: *Molecular Biology and Evolution* 30.3, pp. 669–688 (pages 17, 57, 100, 145).

Xie, W, P Lewis, Y Fan, L Kuo and M Chen (2011). 'Improving marginal likelihood estimation for Bayesian phylogenetic model selection'. In: *Systematic Biology* 60, pp. 150–160 (pages 19, 136, 137).

Yang, Z (1994). 'Maximum likelihood phylogenetic estimation from DNA sequences with variable rates over sites: approximate methods'. In: *Journal of Molecular Evolution* 39.3, pp. 306–314 (pages 54, 103).

Yang, Z and B Rannala (1997). 'Bayesian phylogenetic inference using DNA sequences: a Markov chain Monte Carlo method'. In: *Molecular Biology and Evolution* 14.7, pp. 717–724 (pages 14, 37, 108).

Yang, Z and B Rannala (2006). 'Bayesian estimation of species divergence times under a molecular clock using multiple fossil calibrations with soft bounds'. In: *Molecular Biology and Evolution* 23.1, pp. 212–226 (page 66).

Yang, Z and A Yoder (1999). 'Estimation of the transition/transversion rate bias and species sampling'. In: *Journal of Molecular Evolution* 48.3, pp. 274–283 (page 102).

Yang, Z, N Goldman and A Friday (1995). 'Maximum likelihood trees from DNA sequences: a peculiar statistical estimation problem'. In: *Systematic Biology* 44.3, pp. 384–399 (page 146).

Yang, Z, R Nielsen, N Goldman and A Pedersen (2000). 'Codon-substitution models for hetero-geneous selection pressure at amino acid sites'. In: *Genetics* 155.1, pp. 431–449 (page 111).

Yoder, A and Z Yang (2000). 'Estimation of primate speciation dates using local molecular clocks'. In: *Molecular Biology and Evolution* 17.7, pp. 1081–1090 (page 63).

Ypma, RJF, WM van Ballegooijen and J Wallinga (2013). 'Relating phylogenetic trees to transmission trees of infectious disease outbreaks'. In: *Genetics* 195.3, pp. 1055–1062 (page 43).

Yu, Y, C Than, JH Degnan and L Nakhleh (2011). 'Coalescent histories on phylogenetic networks and detection of hybridization despite incomplete lineage sorting'. In: *Systematic Biology* 60.2, pp. 138–149 (page 120).

Yule, G (1924). 'A mathematical theory of evolution based on the conclusions of Dr. J.C. Willis'. In: *Philosophical Transactions of the Royal Society B: Biological Sciences* 213, pp. 21–87 (pages 36, 38).

Index of authors

Index of subjects